AF358689

Bioactive Natural Products from the Red Sea

Editors

Mostafa Rateb
Usama Ramadan Abdelmohsen

MDPI • Basel • Beijing • Wuhan • Barcelona • Belgrade • Manchester • Tokyo • Cluj • Tianjin

Editors
Mostafa Rateb
School of Computing,
Engineering Physical Sciences
University of the West of
Scotland
Glasgow
United Kingdom

Usama Ramadan Abdelmohsen
Department of Pharmacognosy,
Faculty of Pharmacy
Deraya University
New Minia
Egypt

Editorial Office
MDPI
St. Alban-Anlage 66
4052 Basel, Switzerland

This is a reprint of articles from the Special Issue published online in the open access journal *Marine Drugs* (ISSN 1660-3397) (available at: www.mdpi.com/journal/marinedrugs/special_issues/ Natural_Products_Red_Sea).

For citation purposes, cite each article independently as indicated on the article page online and as indicated below:

LastName, A.A.; LastName, B.B.; LastName, C.C. Article Title. *Journal Name* **Year**, *Volume Number*, Page Range.

ISBN 978-3-0365-1588-5 (Hbk)
ISBN 978-3-0365-1587-8 (PDF)

Bioactive Natural Products from the Red Sea

Contents

marine drugs

Editorial

Bioactive Natural Products from the Red Sea

Mostafa E. Rateb [1,*] and Usama Ramadan Abdelmohsen [2,3,*]

1. School of Computing, Engineering & Physical Sciences, University of the West of Scotland, Paisley PA1 2BE, UK
2. Department of Pharmacognosy, Faculty of Pharmacy, Deraya University, New Minia 61111, Egypt
3. Egypt 11 Department of Pharmacognosy, Faculty of Pharmacy, Minia University, Minia 61519, Egypt
* Correspondence: Mostafa.Rateb@uws.ac.uk (M.E.R.); Usama.ramadan@mu.edu.eg (U.R.A.)

Citation: Rateb, M.E.; Abdelmohsen, U.R. Bioactive Natural Products from the Red Sea. *Mar. Drugs* **2021**, *19*, 289. https://doi.org/10.3390/md 19060289

Received: 10 May 2021
Accepted: 20 May 2021
Published: 21 May 2021

Publisher's Note: MDPI stays neutral with regard to jurisdictional claims in published maps and institutional affiliations.

The marine environment has proven to be a rich source of diverse natural products with relevant activities such as anticancer, anti-inflammatory, antiepileptic, immunomodulatory, antifungal, antiviral, and antiparasitic [1,2]. The global marine pharmaceutical clinical pipeline comprises of 48 compounds originating from different marine invertebrates and marine microorganisms, including 15 approved drugs by the most representative approving agencies, 5 drug candidates in phase III, 12 in phase II, and 16 in phase I of drug development clinical phases [3]. Marine invertebrates and associated microorganisms are capable of synthesizing diverse classes of secondary metabolites and, in some cases, novel chemical leads of which terrestrial counterparts have never been discovered.

The Red Sea is the world's northernmost tropical sea, acting as an inlet of the Indian Ocean, lying between Africa and Asia. It has a surface area of roughly 450,000 km^2 and is approximately 2250 km long, with a maximum depth of around 3000 m. The Red Sea is approximately 5% greater than the world average salinity, due to a high rate of evaporation and a lack of significant rivers or streams draining into it [4]. The Red Sea is a rich and diverse ecosystem due to the 2000 km of coral reef extending along its coastline. It is inhabited by over 1000 invertebrate species and 200 soft and hard corals [5]. Due to this high biodiversity and limited research, the Red Sea is a promising, underexplored habitat for the discovery of new bioactive marine natural products.

This Special Issue contains nine articles, including eight research articles on different topics related to the natural products derived from the Red Sea and one review article. In the following sections, we would provide a short overview of what the reader will find in our Special Issue.

Qader et al. reported the isolation and identification of a new cyclic tripeptide named epicotripeptin, along with four known cyclic dipeptides and one acetamide, derivative from seagrass-associated endophytic fungus *Epicoccum nigrum* M13 recovered from the Red Sea. Moreover, they reported two new compounds, cyclodidepsipeptide phragamide A and trioxobutanamide derivative phragamide B, together with eight known compounds from plant-derived endophyte *Alternaria alternata* 13A collected from a saline lake of the Wadi El Natrun depression in the Sahara Desert. The antimicrobial screening indicated that seven of the tested compounds exhibited considerable (MIC range of 2.5–5 µg/mL) to moderate (10–20 µg/mL) antibacterial effect against the tested Gram-positive strains, and moderate to weak (10–30 µg/mL) antibacterial effect against Gram-negative strains. On the other hand, four of the tested compounds showed considerable antibiofilm effects against biofilm-forming Gram-positive and Gram-negative strains [6].

Cocultivation is a productive technique to trigger microbes' biosynthetic capacity by mimicking the natural habitats' features, principally by competition for food and space and interspecies cross-talks. Alhadrami et al. used coculture of two Red Sea-derived actinobacteria, *Actinokineospora spheciospongiae* strain EG49 and *Rhodococcus* sp. UR59, resulting in the induction of several non-traced metabolites in their axenic cultures, which were detected using LC–HRMS metabolomics analysis. Antimalarial guided isolation of the cocultured fermentation led to the isolation of the angucyclines actinosporins E, H, G, tetragulol, and the anthraquinone capillasterquinone B. The isolated angucycline and anthraquinone compounds exhibited in vitro antimalarial activity and good biding affinity against lysyl-tRNA synthetase (PfKRS1), highlighting their potential to be developed as a new antimalarial structural motif [7].

Tammam et al. were able to isolate six new, and twenty known, steroids from the extract of the soft coral *Sinularia polydactyla*, collected from the Hurghada reef in the Red Sea. They evaluated the cytotoxic, anti-inflammatory, anti-angiogenic, and neuroprotective activities of the isolated compounds and their effect on androgen receptor-regulated transcription in human tumours and non-cancerous cells. Two steroids showed significant cytotoxicity in the low micromolar range against the HeLa and MCF7 cancer cell lines. Additionally, two compounds exhibited neuroprotective activity on neuron-like SH-SY5Y cells [8].

Shaala et al. investigated the actinomycete strain *Streptomyces coelicolor* LY001 recovered from the sponge *Callyspongia siphonella*. The chemical analysis resulted in the isolation of three new natural chlorinated 3-phenylpropanoic acid derivatives, 3-(3,5-dichloro-4-hydroxyphenyl)propanoic acid, 3-(3,5-dichloro-4-hydroxyphenyl)propanoic acid methyl ester, and 3-(3-chloro-4-hydroxyphenyl)propanoic acid, along with 3-phenylpropanoic acid, E-cinnamic acid, and the diketopiperazine alkaloids cyclo(L-Phe-trans-4-OH-L-Pro) and cyclo(L-Phe-cis-4-OH-D-Pro). The compounds demonstrated significant and selective activities towards *Escherichia coli*, *Staphylococcus aureus* and *Candida albicans* [9].

In the article by Abdelhameed et al., the seagrass *Thalassodendron ciliatum* (Forssk.) Den Hartog was investigated. In this work, a new ergosterol derivative named thalassosterol was identified from the methanolic extract of *T. ciliatum*, along with two known sterols, namely ergosterol and stigmasterol. Thalassosterol exhibited significant in vitro antiproliferative potential against the human cervical cancer cell line (HeLa), and human breast cancer (MCF-7) cell lines. Those results aligned with docking studies on the new sterol which explained the possible binding interactions with an aromatase enzyme; this inhibition is beneficial in both cervical and breast cancer therapy [10].

Cocultivation has been known as an effective approach for the enhancement of the production of natural products from bacteria and fungi. In the study by Hifnawy et al. liquid chromatography coupled with high-resolution mass spectrometry-assisted metabolomic profiling of two sponge-associated actinomycetes, *Micromonospora* sp. UR56 and *Actinokineospora* sp. EG49, resulted in the induction of phenazine-derived compounds that were later identified, upon fermentation, as dimethyl phenazine-1,6-dicarboxylate, phenazine-1,6-dicarboxylic acid mono methyl ester (phencomycin, phenazine-1-carboxylic acid (tubermycin), *N*-(2-hydroxyphenyl)-acetamide, and p-anisamide. The antibacterial, antibiofilm, and cytotoxic properties of these metabolites were evaluated via in vitro and docking studies. This study highlighted that microbial cocultivation is an efficient tool for the discovery of new antimicrobial candidates and indicated phenazines as potential lead compounds for further development as antibiotic scaffolds [11].

Abdelhameed et al. reported the isolation of two new compounds: a ceramide, stylissamide A, and a cerebroside, stylissoside A, from the Red Sea sponge *Stylissa carteri*. Metabolomic profiling revealed the presence of diverse secondary metabolites, mainly oleanane-type saponins, phenolic diterpenes, and lupane triterpenes. They also investigated the in vitro cytotoxic activity of the isolated compounds against two human cancer cell lines, MCF-7 and HepG2. Molecular docking experiments showed that both compounds displayed high affinity to the SET protein and inhibitor 2 of protein phosphatase 2A (I2PP2A), a possible mechanism for their cytotoxic activity [12].

El-Kashef et al. investigated the marine-derived fungus *Aspergillus falconensis*, cultivated from sediment collected from the Canyon at Dahab, Red Sea, and reported a yield of two new chlorinated azaphilones, falconensins O and P in addition to four known azaphilone derivatives. Interestingly, replacing NaCl with NaBr induced the accumulation of three additional new azaphilones, falconensins Q−S, including two brominated derivatives along with three known analogues. Some of the tested compounds showed NF-κB inhibitory activity against the triple-negative breast cancer cell line MDA-MB-231. This study highlights the OSMAC approach as a successful approach to increase the diversity of secondary metabolites from marine fungi [13].

El-Hossary et al. provided a comprehensive review that compares the natural products recovered from the Red Sea in terms of ecological role and pharmacological activities. In this review, which covers the literature to the end of 2019, they summarized the diversity of bioactive secondary metabolites derived from Red Sea micro- and macro-organisms, and discuss their biological potential whenever applicable. Moreover, the diversity of the Red Sea organisms is highlighted, as well as genomic potential [14].

Finally, we could see that Red Sea derived natural products were diverse in terms of chemical scaffolds as well as pharmacological activities.

As guest editors, we appreciate the efforts provided by all of the authors who contributed their excellent results to this Special Issue, all of the reviewers who carefully evaluated the submitted manuscripts, and the editorial boards of Marine Drugs for their support and kind help.

Funding: This research received no external funding.

Conflicts of Interest: The authors declare no conflict of interest.

References

1. Carroll, A.R.; Copp, B.R.; Davis, R.A.; Keyzers, R.A.; Prinsep, M.R. Marine natural products. *Nat. Prod. Rep.* **2020**, *37*, 175–223. [CrossRef]
2. Carroll, A.R.; Copp, B.R.; Davis, R.A.; Keyzers, R.A.; Prinsep, M.R. Marine natural products. *Nat. Prod. Rep.* **2019**, *36*, 122–173. [CrossRef] [PubMed]
3. Available online: https://www.marinepharmacology.org/ (accessed on 14 May 2021).
4. Available online: https://www.worldatlas.com/seas/red-sea.html (accessed on 8 May 2021).
5. Available online: https://www.newworldencyclopedia.org/entry/Red_Sea (accessed on 8 May 2021).
6. Qader, M.M.; Hamed, A.A.; Soldatou, S.; Abdelraof, M.; Elawady, M.E.; Hassane, A.S.I.; Belbahri, L.; Ebel, R.; Rateb, M.E. Antimicrobial and Antibiofilm Activities of the Fungal Metabolites Isolated from the Marine Endophytes *Epicoccum nigrum* M13 and *Alternaria alternata* 13A. *Mar. Drugs* **2021**, *19*, 232. [CrossRef] [PubMed]
7. Alhadrami, H.A.; Thissera, B.; Hassan, M.H.A.; Behery, F.A.; Ngwa, C.-J.; Hassan, H.M.; Pradel, G.; Abdelmohsen, U.R.; Rateb, M.E. Bio-Guided Isolation of Antimalarial Metabolites from the Coculture of Two Red Sea Sponge-Derived *Actinokineospora* and *Rhodococcus* spp. *Mar. Drugs* **2021**, *19*, 109. [CrossRef] [PubMed]
8. Tammam, M.A.; Rárová, L.; Kvasnicová, M.; Gonzalez, G.; Emam, A.M.; Mahdy, A.; Strnad, M.; Ioannou, E.; Roussis, V. Bioactive Steroids from the Red Sea Soft Coral *Sinularia polydactyla*. *Mar. Drugs* **2020**, *18*, 632. [CrossRef] [PubMed]
9. Shaala, L.A.; Youssef, D.T.A.; Alzughaibi, T.A.; Elhady, S.S. Antimicrobial Chlorinated 3-Phenylpropanoic Acid Derivatives from the Red Sea Marine Actinomycete *Streptomyces coelicolor* LY001. *Mar. Drugs* **2020**, *18*, 450. [CrossRef] [PubMed]
10. Abdelhameed, R.F.A.; Habib, E.S.; Goda, M.S.; Fahim, J.R.; Hassanean, H.A.; Eltamany, E.E.; Ibrahim, A.K.; AboulMagd, A.M.; Fayez, S.; Abd El-kader, A.M.; et al. Thalassosterol, a New Cytotoxic Aromatase Inhibitor Ergosterol Derivative from the Red Sea Seagrass *Thalassodendron ciliatum*. *Mar. Drugs* **2020**, *18*, 354. [CrossRef] [PubMed]
11. Hifnawy, M.S.; Hassan, H.M.; Mohammed, R.; Fouda, M.M.; Sayed, A.M.; Hamed, A.A.; AbouZid, S.F.; Rateb, M.E.; Alhadrami, H.A.; Abdelmohsen, U.R. Induction of Antibacterial Metabolites by Co-Cultivation of Two Red-Sea-Sponge-Associated Actinomycetes *Micromonospora* sp. UR56 and *Actinokinespora* sp. EG49. *Mar. Drugs* **2020**, *18*, 243. [CrossRef] [PubMed]
12. Abdelhameed, R.F.A.; Habib, E.S.; Eltahawy, N.A.; Hassanean, H.A.; Ibrahim, A.K.; Mohammed, A.F.; Fayez, S.; Hayallah, A.M.; Yamada, K.; Behery, F.A.; et al. New Cytotoxic Natural Products from the Red Sea Sponge *Stylissa carteri*. *Mar. Drugs* **2020**, *18*, 241. [CrossRef]
13. El-Kashef, D.H.; Youssef, F.S.; Hartmann, R.; Knedel, T.; Janiak, C.; Lin, W.; Reimche, I.; Teusch, N.; Zhen Liu, Z.; Proksch, P. Azaphilones from the Red Sea Fungus *Aspergillus falconensis*. *Mar. Drugs* **2020**, *18*, 204. [CrossRef]
14. El-Hossary, E.M.; Abdel-Halim, M.; Ibrahim, E.S.; Pimentel-Elardo, S.M.; Nodwell, J.R.; Handoussa, H.; Abdelwahab, M.F.; Holzgrabe, U.; Abdelmohsen, U.R. Natural Products Repertoire of the Red Sea. *Mar. Drugs* **2020**, *18*, 457. [CrossRef]

 *marine drugs*

Review

Natural Products Repertoire of the Red Sea

Ebaa M. El-Hossary [1], Mohammad Abdel-Halim [2], Eslam S. Ibrahim [3,4],
Sheila Marie Pimentel-Elardo [5], Justin R. Nodwell [5], Heba Handoussa [6],
Miada F. Abdelwahab [7], Ulrike Holzgrabe [8,*] and Usama Ramadan Abdelmohsen [7,9,*]

[1] National Centre for Radiation Research & Technology, Egyptian Atomic Energy Authority,
    Ahmed El-Zomor St. 3, El-Zohoor Dist., Nasr City, Cairo 11765, Egypt; ebaa.elhossary@eaea.org.eg
[2] Department of Pharmaceutical Chemistry, Faculty of Pharmacy and Biotechnology,
    German University in Cairo, Cairo 11835, Egypt; mohammad.abdel-halim@guc.edu.eg
[3] Department of Microbiology and Immunology, Faculty of Pharmacy, Cairo University, Cairo 11562, Egypt;
    eslam.ebrahim@pharma.cu.edu.eg
[4] Institute for Molecular Infection Biology, University of Würzburg, Josef-Schneider-Strasse 2/Bau D15,
    97080 Würzburg, Germany
[5] Department of Biochemistry, University of Toronto, MaRS Centre West, 661 University Avenue,
    Toronto, ON M5G 1M1, Canada; sheila.elardo@utoronto.ca (S.M.P.-E.); justin.nodwell@utoronto.ca (J.R.N.)
[6] Department of Pharmaceutical Biology, Faculty of Pharmacy and Biotechnology,
    German University in Cairo, Cairo 11835, Egypt; heba.handoussa@guc.edu.eg
[7] Department of Pharmacognosy, Faculty of Pharmacy, Minia University, Minia 61519, Egypt;
    mayada.mohamed2@mu.edu.eg
[8] Institute for Pharmacy and Food Chemistry, University of Würzburg, Am Hubland,
    97074 Würzburg, Germany
[9] Department of Pharmacognosy, Faculty of Pharmacy, Deraya University, Universities Zone,
    P.O. Box 61111 New Minia City, Minia 61519, Egypt
* Correspondence: ulrike.holzgrabe@uni-wuerzburg.de (U.H.); usama.ramadan@mu.edu.eg (U.R.A.)

Received: 2 August 2020; Accepted: 2 September 2020; Published: 4 September 2020

Abstract: Marine natural products have achieved great success as an important source of new lead compounds for drug discovery. The Red Sea provides enormous diversity on the biological scale in all domains of life including micro- and macro-organisms. In this review, which covers the literature to the end of 2019, we summarize the diversity of bioactive secondary metabolites derived from Red Sea micro- and macro-organisms, and discuss their biological potential whenever applicable. Moreover, the diversity of the Red Sea organisms is highlighted as well as their genomic potential. This review is a comprehensive study that compares the natural products recovered from the Red Sea in terms of ecological role and pharmacological activities.

Keywords: Red Sea; marine natural products; marine organisms; biodiversity; marine metagenomics; bioactivity

---

1. Introduction

Around 24 million years ago, the separation of the African and Arabian plates created the Red Sea [1]. Since then, the Red Sea has been characterized by exclusive assets such as location, relatively high temperature and relatively young geologic age. In the north, the Suez Canal connects the Mediterranean Sea with the Red Sea. While in the south, the Red Sea is connected to the Gulf of Aden and the Indian Ocean through Bab El-Mandeb Strait (Figure 1). The Red Sea exhibits an enormous diversity of life in all domains of life [2]. However, its biosphere is still underexplored and poorly understood [3,4]. Since 2000, no less than 58 new endemic species were identified in the Red Sea, for example, *Inermonephtys aramco* from the southern region [5,6]. These new species expanded our understanding of the previously discovered organisms such as *Aglaophamus lobatus*, *Aglaophamus* cf. *verrilli* and *Micronephthys stammeri* [6]. The Red Sea acts as a unique source of biological diversity by exporting species and genetic lineages for the global marine biodiversity patterns [7–9].

Figure 1. A satellite image of the Red Sea. The marine organisms were collected from Egypt (El Gouna, Hurghada, Ras Gharib, Ras Muhammad, Safaga, Gulf of Aqaba, and Sharm El-Sheikh), Saudi Arabia (Hakel area, Jazan, Obhur, Al-Lith, Jeddah, Salman Gulf and Yanbu), Israel (Gulf of Eilat), Eritrea (Dahlak archipelago and Massawa), Djibouti (Ardoukoba), Jordan (Aqaba) and Yemen (Hanish Islands).

The Red Sea biosphere is divided into macrofauna (organisms larger than 0.5 or 1 mm in diameter) [10], meiofauna (organisms less than 0.5 mm in diameter) [11] and microorganisms [12–14]. In general, biodiversity may be influenced by nutrients localization [15], sediment particle size ranges [15], salinity, degree of oxygenation [16], depth (pressure) [17,18], temperature [19], organic matter, hydrodynamics [20], light conditions [21], as well as any natural or anthropogenic changes in these abiotic factors [22]. The Red Sea has an oligotrophic nature, which affects the biodiversity due to the deficiency of major nutrients, including nitrate, ammonium, phosphate and silica [21]. The most oligotrophic location is the central northern region owing to weak mixing and limited nutrient flux [2,23]. However, in the south, nutrient-rich flux from the Indian Ocean compensates for oligotrophy [24,25]. It is important to notice that the marine ecosystem has a capacity to equilibrate different factors including the major nutrients [26]. A salinity gradient is observed from north to south (42 to 37 psu) [1,27]. Surface temperatures vary from 24 to 35 °C in spring to summer, respectively [27]. Due to abiotic factors, the central Red Sea holds the highest species richness of marine organisms [27].

The anthropogenic activity of SCUBA diving (self-contained underwater breathing apparatus) was reported to negatively affect the corals which may lead to imbalances in the associated corallivorous and herbivorous fish [28]. The sedimentation rate changes depending on the diving intensities [28]. Moreover, the anthropogenic activities are related to warming of the Red Sea which was initiated in the mid-90s, with an 0.7 °C increment after 1994 [29]. Since the Red Sea is a bio-source for the global marine biodiversity, artificial conditions might be used to simulate this dynamicity to scrutinize the anthropogenic negative effects [30].

On the other hand, marine environment has proven to be a very rich source of diverse natural products with biological activities such as anticancer, antibacterial, antifungal, antiviral and antiparasitic [31–34]. The global marine pharmaceutical clinical pipeline comprises about 30 compounds originating from different marine invertebrates and marine microorganisms. These 30 compounds include 8 approved drugs by the most representative approving agencies and 22 drug candidates in phase I, II or III of drug development clinical phases [35–39]. Marine invertebrates and associated microorganisms are capable of synthesizing diverse classes of secondary metabolites and, in some cases, novel chemical leads that have never been discovered in terrestrial counterparts.

In March 2017, a review article containing a listing of natural products, isolated from Red Sea marine organisms, was published [40]. This published review covered the literature until the year 2014, and presented the chemical structures of 435 natural products together with their biological activities whenever applicable. The natural products mentioned in this published review were categorized according to their chemical classes, including terpenes, alkaloids, sterols, steroidal glycosides and other metabolites.

In our review, we conducted a comprehensive literature search covering the time period until December 2019, with the inclusion of extra five years (from 2015 to 2019). We found that a total of 677 marine natural products were isolated form Red Sea marine organisms until the end of 2019. Publications that describe extracts and structurally uncharacterized marine natural products have been excluded from our review. The structures and biological activities of selected 111 marine natural products are illustrated in this review, and categorized according to their isolation source (marine organism), including microorganisms, invertebrates, algae and sea grasses. The presented 111 marine natural products were selected due to their potent biological activities. These selected compounds exhibited a wide range of potent biological activities, such as antioxidant, anticonvulsant, anticancer and anti-infective activities. In the Supplementary data, a list of the remaining 566 marine natural products from the Red Sea (with lower or no biological activities) is presented in Table S1.

The review illustrates also the geographical distribution of the collection sites of the marine organisms. Marine organisms, which are mentioned in this review, were collected from the Red Sea coasts of seven countries (Figure 2). These countries include Egypt, Saudi Arabia, Israel, Eritrea, Djibouti, Jordan and Yemen. The locations of collection of each marine organism are presented in Table S2 in the Supplementary data. The numbers of the collected marine micro- and macro-organisms from the Red Sea are presented in Figure 3.

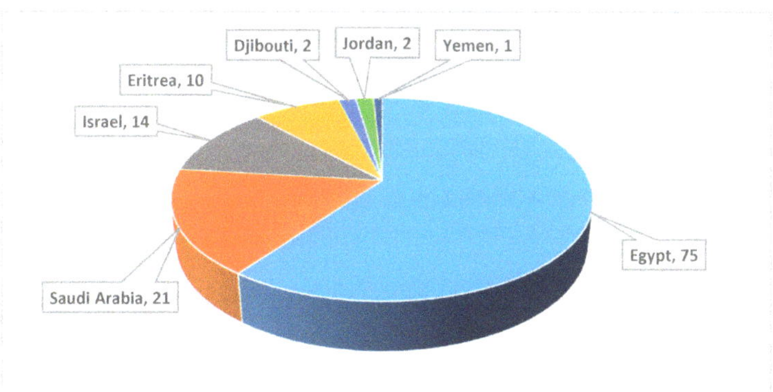

Figure 2. Numbers of marine organisms according to the location of collection in the Red Sea.

Taxonomy of the marine organisms collected from the Red Sea is also presented in Table S3 in the Supplementary data. The correlation between the numbers and chemical classes of the isolated compounds and their isolation sources (marine organisms) is shown in Figure 4. Moreover, a comprehensive study on the genomic potential of the Red Sea organisms is included in the review. To the best of our knowledge, this is the first comprehensive review that presents and discusses the marine natural products of the Red Sea until the end of 2019, with regard to their chemistry, biological activities, biodiversity of the collected marine organisms, geographical distribution of the locations of collection of the marine organisms and the genomic potential of the marine organisms of the Red Sea.

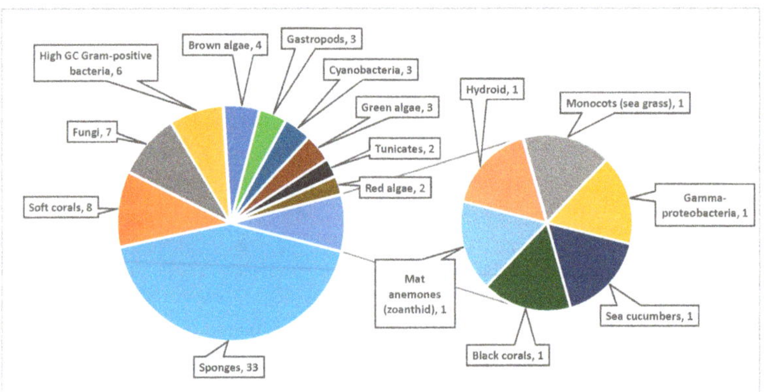

Figure 3. Numbers of marine organisms collected from the Red Sea.

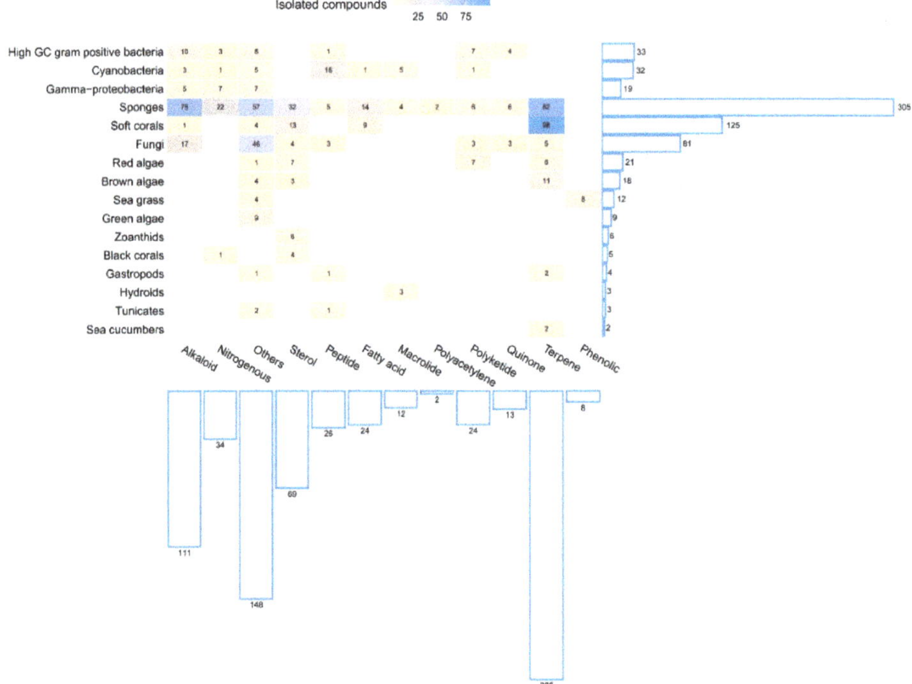

Figure 4. Numbers and chemical classes of the isolated compounds and their isolation sources (marine organisms).

2. Marine Natural Products Isolated from Marine Microorganisms

2.1. Marine Bacteria

Nuclear magnetic resonance- and mass spectrometry-guided fractionation of the extract of *Okeania* sp. marine cyanobacterium, collected from the Saudi coast from the Algetah Alkabira reef near

Jeddah, afforded the bioactive metabolites lyngbyabellin O (**1**), lyngbyabellin P (**2**), lyngbyabellin G (**3**) and dolastatin 16 (**4**) (Figure 5). The identified compounds **1**, **2**, **3** and **4** were evaluated for their antifouling activity, using *Amphibalanus amphitrite* larvae. Compounds **1**, **2** and **3** displayed remarkable antifouling activity with EC_{50} values of 0.38, 0.73 and 7.4 μM, respectively. Compound **4**, which is reported as a potent antifouling agent, was the most active one with EC_{50} value of 0.09 μM. On the other hand, compound **2** was cytotoxic to MCF7 cells with GI_{50} value of 9 μM, while compound **3** showed no cytotoxicity to MCF7 cells (GI_{50} > 160 μM). Lyngbyabellin G (**3**) exhibited cytotoxicity on MCF7, H460 and Neuro-2a cells with GI_{50} values of 120, 2.2 and 4.8 μM, respectively [41–43]. Okino and co-workers isolated a cytotoxic cyanobactin, wewakazole B (**5**), from the brownish-red filamentous cyanobacterium *Moorea producens* (formerly *Lyngbya majuscula*), which was collected near Jeddah, Saudi Arabia (Figure 5). Wewakazole B (**5**) exhibited cytotoxic activity towards human MCF7 breast cancer cells (IC_{50} = 0.58 μM) and human H460 lung cancer cells (IC_{50} = 1 μM). Unlike what was reported for many cyanobactins, wewakazole B (**5**) exhibited no metal-binding activity at 89 μM, excluding its function as siderophore [44].

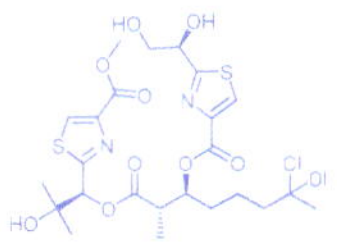

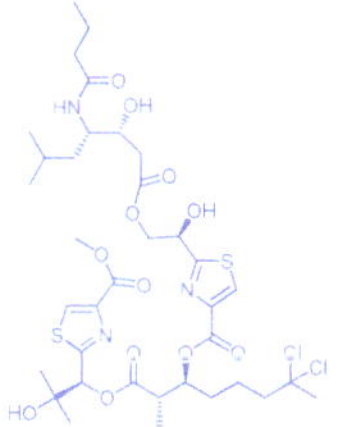

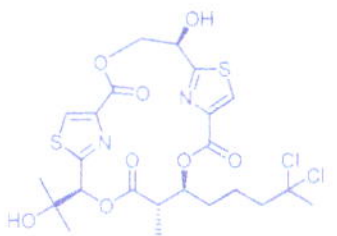

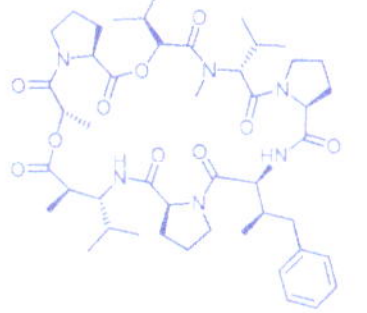

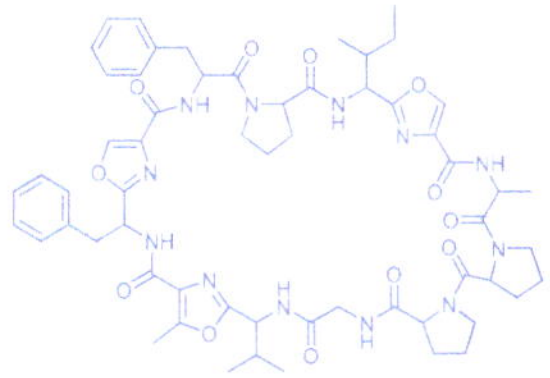

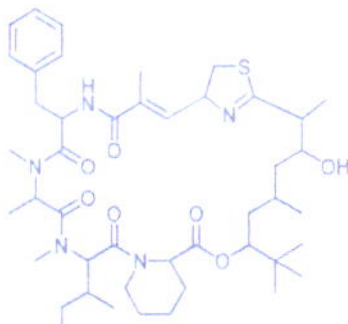

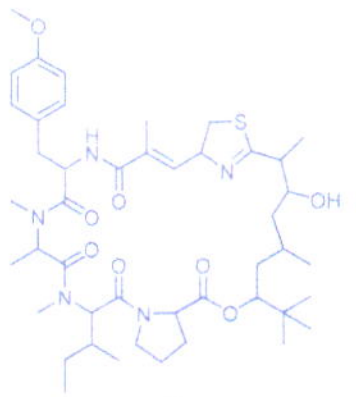

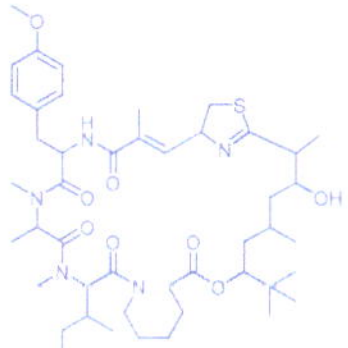

Figure 5. Chemical structures of lyngbyabellin O (**1**), lyngbyabellin P (**2**), lyngbyabellin G (**3**), dolastatin 16 (**4**), wewakazole B (**5**), apratoxin H (**6**), apratoxin A sulfoxide (**7**), grassypeptolide D (**8**) and grassypeptolide E (**9**).

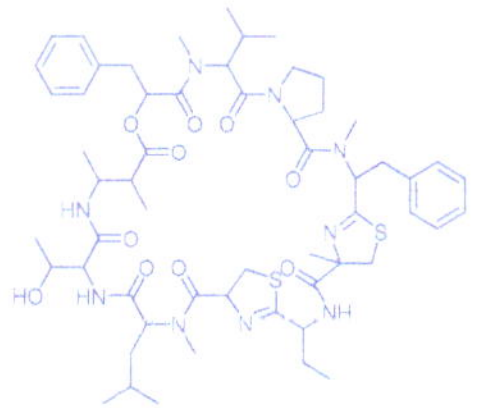

A cultivated *Moorea producens*, collected from the Nabq Mangroves in the Gulf of Aqaba near Sharm el-Sheikh, Egypt, was the source of the cyclic depsipeptides apratoxin H (**6**) and apratoxin A sulfoxide (**7**) (Figure 5). Apratoxin H (**6**) showed potent cytotoxicity to human NCI-H460 lung cancer cells with IC_{50} value of 3.4 nM, while apratoxin A sulfoxide (**7**) was 26-fold less active than apratoxin H (**6**) (IC_{50} = 89.9 nM). It is worth mentioning that the S-oxide function in apratoxin A sulfoxide (**7**) led to about 36-fold reduction in potency in comparison with its parent compound apratoxin A, the well-studied analog as anticancer agent [45]. Two macrocyclic depsipeptides, namely grassypeptolide D (**8**) and grassypeptolide E (**9**), were isolated from the marine cyanobacterium *Leptolyngbya* sp., which was collected from the *SS Thistlegorm* shipwreck in the Red Sea (Figure 5). Grassypeptolide D (**8**) and grassypeptolide E (**9**) exhibited potent cytotoxicity at nanomolar concentrations to HeLa cells (IC_{50} = 335 and 192 nM, respectively) and mouse neuro-2a blastoma cells (IC_{50} = 599 and 407 nM, respectively) [46].

The actinosporin analogues actinosporin C (**10**) and actinosporin D (**11**) were isolated from the calcium alginate beads culture of sponge-associated *Actinokineospora* sp. strain EG49, which was cultivated from the sponge *Spheciospongia vagabunda* (Figure 6). Actinosporin C (**10**) and actinosporin D (**11**) exhibited significant antioxidant activity at a concentration of at 1.25 µM, and protective capacity from the genomic damage induced by hydrogen peroxide in the human promyelocytic (HL-60) cell line. Compounds **10** and **11** did not show any significant cytotoxicity against the mammalian cells HL-60 after 4 h at concentrations of up to 50 µM, while after 24 h, significant cytotoxicity for compound **10** was observed at a concentration of 50 µM [47]. In another study, compounds **10** and 11 were also isolated from *Actinokineospora spheciospongiae* sp. nov, which was isolated from the sponge *Spheciospongia vagabunda* that was collected from offshore Ras Mohamed, Egypt [48]. Saadamycin (**12**) was isolated from the endophytic *Streptomyces* sp. Hedaya48, isolated from the marine sponge *Aplysina fistularis*, which was collected from the Egyptian coast of Sharm El-Sheikh (Figure 6). Saadamycin (**12**) displayed significant antifungal activity against nine dermatophytes and other clinical fungi, with IC_{50} values of 1–5.16 µg/mL [49]. Chromatographic separation of the organic extract of the marine cyanobacterium *Moorea producens*, collected from the Saudi coast near Obhur, led to the isolation of lyngbyatoxin A (**13**) and debromoaplysiatoxin (**14**) (Figure 6) [50]. In a previous report, compound **14** was also isolated form the same marine cyanobacterium, which was collected near Jeddah [51]. Compounds **13** and **14** exhibited potent antiproliferative activity against HeLa cancer cells with IC_{50} value of 9.2 nM and 3.03 µM [50]. The (1*H*)-pyrazinone alkaloid 3,6-diisobutyl-2(1*H*)-pyrazinone (**15**) was isolated from the tunicate-derived actinomycete *Streptomyces* sp. Did-27, collected from a location near Obhur, Saudi Arabia (Figure 6). The alkaloid 15 displayed selective cytotoxicity against HCT-116 cell line with IC_{50} value of 1.5 µg/mL [52]. The four bioactive metabolites chrysophanol 8-methyl ether (**16**), asphodelin (**17**), justicidin B (**18**) and ayamycin (**19**) were isolated from the actinomycete *Nocardia* sp. ALAA 2000, derived from the marine red alga *Laurencia spectabilis*, which was collected from the Egyptian coast in Ras-Gharib (Figure 6). Compounds **16–19** showed significant antimicrobial potential against eight pathogenic bacterial strains, including Gram-positive and Gram-negative bacteria, as well as six fungal strains, with MIC values ranging from 0.1 to 10 µg/mL [53]. Fridamycin H (**20**) was isolated from the elicited marine sponge-derived bacterium *Actinokineospora spheciospongiae* sp. nov., collected from offshore Ras Mohamed, Egypt. Fridamycin H (**20**) exhibited antitrypanosomal activity (against *Trypanosoma brucei* strain TC221) after 48 and 72 h, with IC_{50} values of 7.18 and 3.35 µM, respectively. Furthermore, compound **20** has no cytotoxic activity against J774.1 macrophages (IC_{50} > 200 µM) [48].

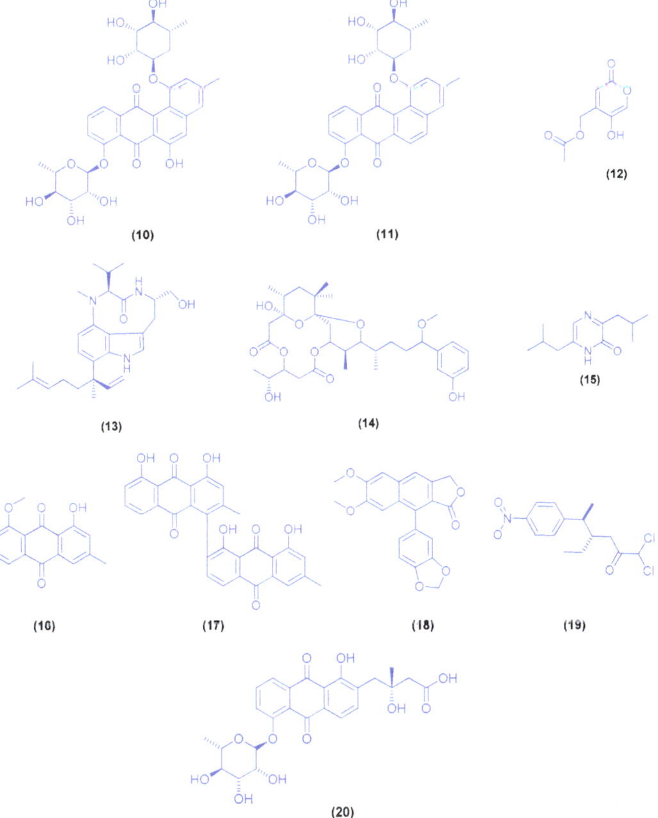

Figure 6. Chemical structures of actinosporin C (**10**), actinosporin D (**11**), saadamycin (**12**), lyngbyatoxin A (**13**), debromoaplysiatoxin (**14**), 3,6-diisobutyl-2(1*H*)-pyrazinone (**15**), chrysophanol 8-methyl ether (**16**), asphodelin (**17**), justicidin B (**18**), ayamycin (**19**) and fridamycin H (**20**).

2.2. Marine Fungi

The xanthone derivative AGI-B4 (**21**) and the cyclic depsipeptide scopularide A (**22**) were isolated from solid rice cultures of the marine-derived fungus *Scopulariopsis* sp., which was obtained from the hard coral *Stylophora* sp. (Figure 7). Compounds **21** and **22** showed significant cytotoxicity against L5178Y mouse lymphoma cells with IC$_{50}$ values of 1.5 and 1.2 µM, respectively, which were lower than that of the positive control kahalalide F (IC$_{50}$ = 4.3 µM) [54].

Figure 7. Chemical structures of AGI-B4 (**21**) and scopularide A (**22**).

3. Marine Natural Products Isolated from Marine Invertebrates

3.1. Sponges

Several cyclic peroxide norterpenoids have been isolated from the sponge *Diacarnus erythraeanus*, which was obtained from Elfanadir (Hurghada Coast, Egypt). These metabolites include norsesterterpene derivatives **23**, **24** and (−)-muqubilin A (**25**) (Figure 8). The three compounds displayed potent growth inhibitory activity against seven different human cancer cell lines, the Hs683 oligodendroglioma, MCF-7 breast, PC-3 prostate, U373 and U251 glioblastoma, SKMEL-28 melanoma and A549 non-small cell lung cancer cells, with a range of IC_{50} values from 1 to 8 μM. Further studies on (−)-muqubilin A (**25**) revealed that it possesses cytotoxicity without selectivity between normal and cancer cells. It induces reactive oxygen species (ROS) production in cancer cells; it is less likely that (−)-muqubilin A (**25**) exerts its cytotoxic action via stimulation of pro-apoptotic processes. It is worth mentioning that, the presence of the free carboxylic acid group in the structure of these metabolites may be critical for their growth inhibitory activity in various cancer cell lines [55]. Sipholenol A (**26**) is a sipholane triterpenoid that was isolated from the marine sponge *Callyspongia siphonella* (formerly known as *Siphonochalina siphonella*), which was collected from Hurghada, at the Egyptian coast (Figure 8) [56]. In another three published studies, compound **26** was also isolated from the marine sponge *Callyspongia siphonell*, collected from Sharm Obhur in Jeddah, Saudi Arabia [57] and from Hurghada in Egypt [58,59]. It was shown that Sipholenol A (**26**)—at non-cytotoxic concentration (5 μM)- could increase the sensitivity of the resistant cervical adenocarcinoma KB-C2 cells by 16 times towards colchicine through the reversal of P-Glycoprotein-mediated MDR. Compound (**26**) did not affect the IC_{50} value of colchicine to the parent cell line KB-3-1(which do not express P-gp) [56]. In a later report, the same research group isolated from the marine sponge *Callyspongia (Siphonochalina) siphonell* other sipholane triterpenoids with the perhydrobenzoxepine nucleus including sipholenone E (**27**), Sipholenol L (**28**) and siphonellinol D (**29**) (Figure 8) [60]. Compound **28** was also isolated from the marine sponge *Siphonochalina siphonella*, collected from Sharm Obhur, Jeddah, Saudi Arabia [57]. Compound **29** was also isolated from the sponge *Siphonochalina siphonella*, which was collected near Hurghada, Egypt [58]. The sipholane triterpenoid **27** was shown to be a superior compound compared to sipholenol A (**26**), in reversing the P-gp-mediated multidrug resistance. Compounds **28** and **29** were comparable to sipholenol A (**26**) in potency. Compounds **26–29** were found to be non-toxic to the human epidermoid cancer cells KB-3-1 and KB-C2, with IC_{50} value higher than 50 μM for both cell lines [60]. Latrunculin A (**30**), latrunculin B (**31**) and 16-*epi*-latrunculin B (**32**) were isolated from the marine sponge *Negombata magnifica*, collected from the coast of Eilat (Figure 8). Latrunculins are macrolides that were reported to cause unique microfilament inhibition due to one-to-one reversible complex with monomeric actin, disrupting its polymerization and consequently its functions within the cell. Latrunculin A (**30**) and latrunculin B (**31**) were able to show activity in the actin-disruption assay in concentrations of 0.5–1 and 0.5–10 μg/mL, respectively (these are the concentrations where microfilament loss is detected). Compound **32** was less active than compounds **30** and **31**, and showed activity at 5–10 μg/mL [61]. Latrunculin B (**31**) inhibited also the migration of B16B15b tumor cell in a wound-healing assay at 1 μM [62].

Compounds **33** and **34** were isolated as a ceramide mixture from the marine sponge *Negombata corticata*, which was collected from Safaga in Egypt (Figure 9). The mixture was evaluated for its in vitro anticonvulsant activity in the pentylenetetrazole-induced seizure model. It showed antiepileptic effect comparable to that of the reference drug diazepam [63]. The bis-1-oxaquinolizidine *N*-oxide alkaloid araguspongine C (**35**) was isolated from the sponge *Xestospongia exigua*, which was obtained from vertical reef slopes at Bayadha, north of Jeddah on the Saudi Arabian coast (Figure 9). Araguspongine C (**35**) was reported to be active against *Plasmodium falciparum* African (D6) clone with an IC_{50} value of 0.67 μg/mL and selectivity index >7.1, and against *P. falciparum* Indochina (W2) clone with IC_{50} value of 0.28 μg/mL and selectivity index >17. Additionally, araguspongine C (**35**) was also shown to possess high activity against the H37Rv strain of *Mycobacterium tuberculosis* with MIC value of 3.94 μM [64].

The norsesterterpene peroxide acids muqubilin (**36**) and sigmosceptrellin B (**37**) were isolated from the marine sponge *Diacarnus erythraeanus*, which was obtained from the Egyptian coast northeast of Hurghada (Figure 9) [65]. In another published study, muquabilin (**36**) was also identified and isolated from the sponge *Diacarnus erythraenus*, which was collected from El Qusier, 120 km south of Hurghada, Egypt [66]. Sigmosceptrellin B (**37**) was also identified and isolated in another two studies from the same marine sponge, which was collected from Hurghada [55,67]. Muqubilin (**36**) and sigmosceptrellin B (**37**) could show potent in vitro antiparasitic activity against *Toxoplasma gondii* at a concentration of 0.1 μM without significant toxicity to HFF or L929 cells. Sigmosceptrellin B (**37**) exhibited also a significant in vitro antimalarial activity against *Plasmodium falciparum* (D6 and W2 clones) with IC_{50} values of 1.2 and 3.4 μg/mL, respectively. The functional group homology of compounds **36** and **37** with artemisinin suggests that the bioactivity may be due to the peroxide moieties [65]. The sponge *Theonella swinhoei*, which was collected in Hurghada, was reported to be the source of swinholide I (**38**) (a macrolide with a symmetric 44-membered dilactone skeleton) and hurghadolide A (**39**) (asymmetric 42-membered dilactone) (Figure 9). Both compounds showed in vitro cytotoxicity against human colon adenocarcinoma (HCT-116) with IC_{50} values of 5.6 and 365 nM, respectively. However, on measuring the actin microfilament-disrupting effects, hurghadolide A (**39**) showed 10 times higher potency compared to swinholide I (**38**) (these compounds caused disruption of the actin cytoskeleton at concentrations of 7.3 and 70 nM, respectively). This suggested that swinholide I (**38**) has higher selectivity for cancer cells and there might be other mechanisms that initiate cytotoxicity in cancer cells. The nonspecific toxicity of hurghadolide A (**39**) might be also indicated by its higher potency against *Candida albicans* (MIC = 31.1 μg/mL for compound **39** vs. 500 μg/mL for compound **38**) [68].

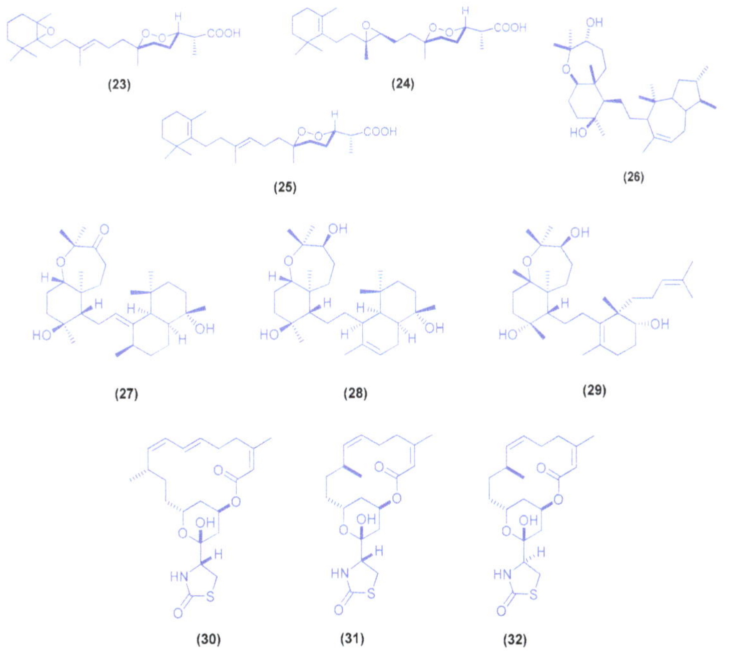

Figure 8. Chemical structures of compound **23**, compound **24**, (−)-muqubilin A (**25**), sipholenol A (**26**), sipholenone E (**27**), sipholenol L (**28**), siphonellinol D (**29**), latrunculin A (**30**), latrunculin B (**31**) and 16-*epi*-latrunculin B (**32**).

The C22-polyacetylenic alcohols callyspongenol A (**40**) and callyspongenol B (**41**) together with dehydroisophonochalynol (**42**) were isolated from the organic extract of the sponge *Callyspongia* sp., which was collected from the Hurghada, Egypt (Figure 10). Compounds **40**, **41** and **42** displayed cytotoxicity against P388 and HeLa cells with a range of IC$_{50}$ values ranging from 2.2 to 5.1 µg/mL [69]. A cytotoxic polyacetylenic amide, callyspongamide A (**43**), was isolated from the marine sponge *Callyspongia fistularis* collected from Hurghada, Egypt (Figure 10). Callyspongamide A (**43**) exhibited a cytotoxic activity against HeLa cells with an IC$_{50}$ value of 4.1 µg/mL [70]. The two indole alkaloids hyrtioerectine B (**44**) and hyrtioerectine C (**45**) were isolated from the marine sponge *Hyrtios erectus*, which was collected from Hurghada, Egypt (Figure 10). The alkaloids 44 and 45 exhibited cytotoxic activity against HeLa cells with IC$_{50}$ values of 5 and 4.5 µg/mL, respectively [71]. The dichloromethane extract of the sponge *Dysidea cinereal*, which was collected near Ras Zaatir in the Gulf of Eilat, yielded the avarol and avarone derivatives avarone A (**46**), 3′,6′-dihydroxyavarone (**47**), avarol C (**48**) and avarone E (**49**) (Figure 10). The avarone derivatives 47 and 49 showed activity as inhibitors of HIV-1 reverse transcriptase, with IC$_{50}$ values of 5 and 1 µg/mL, respectively. The cytotoxicity of the compounds was also measured against P388 mouse leukemia cells. The IC$_{50}$ values were found to be 0.6, 1.2 and <0.6 µg/mL for compounds **46–48**, respectively [72].

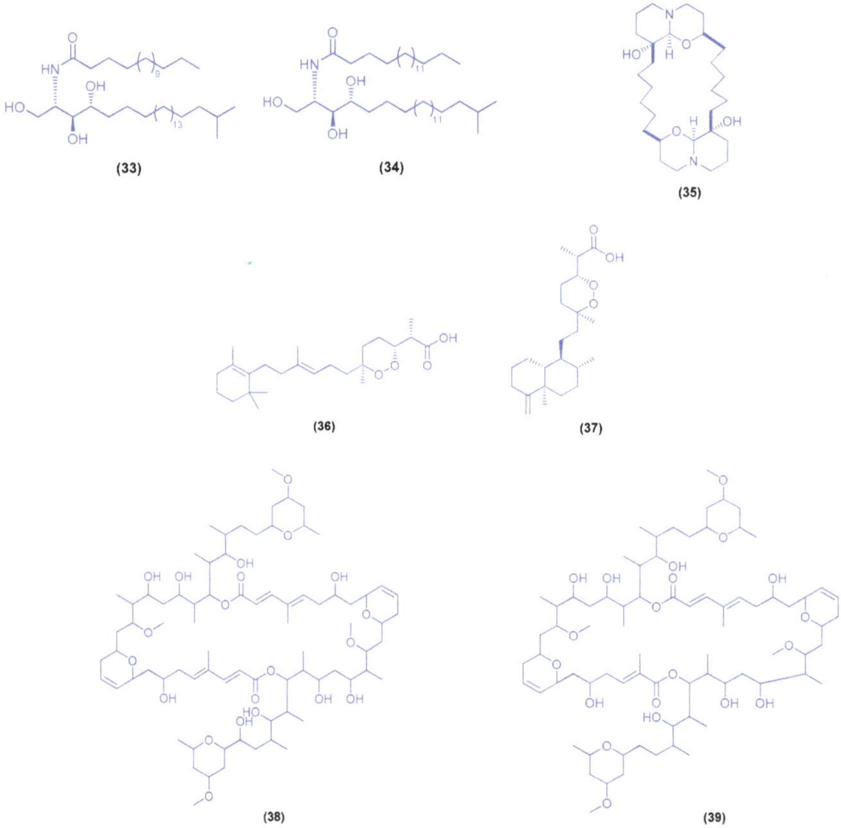

Figure 9. Chemical structures of compounds 33, compound **34**, araguspongine C (**35**), muqubilin (**36**), sigmosceptrellin B (**37**), swinholide I (**38**), hurghadolide A (**39**).

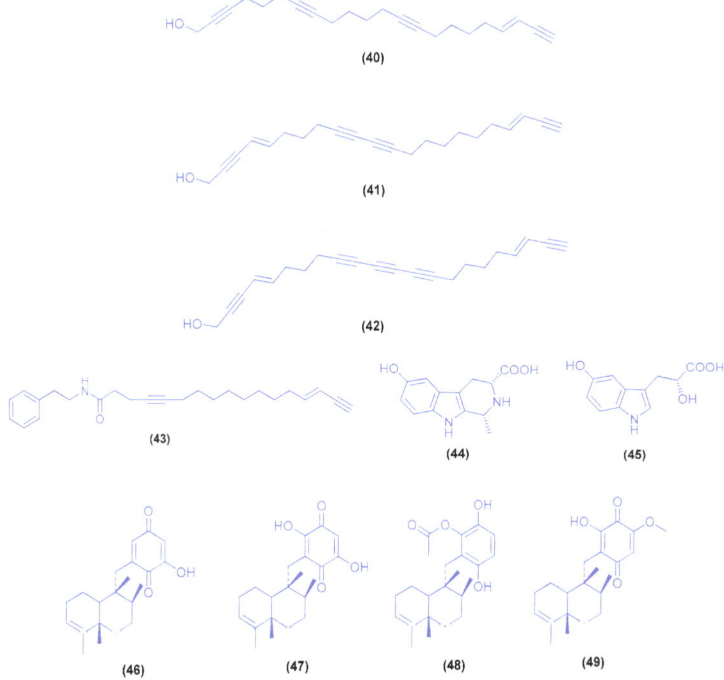

Figure 10. Chemical structures of callyspongenol A (**40**), callyspongenol B (**41**), dehydroisophonochalynol (**42**), callyspongamide A (**43**), hyrtioerectine B (**44**), hyrtioerectine C (**45**), avarone A (**46**), 3′,6′-dihydroxyavarone (**47**), avarol C (**48**) and avarone E (**49**).

In 1993, a bioassay-guided fractionation of an organic extract of the sponge *Toxiclona toxius* yielded the hexaprenoid hydroquinones toxiusol (**50**), shaagrockol C (**51**), toxicol A (**52**) and toxicol B (**53**) (Figure 11) [73]. In the same year, another study published the isolation of toxiusol (**50**), toxicol A (**52**) and toxicol B (**53**) form the marine sponge *Toxiclona toxius* [74]. One year earlier, shaagrockol C (**51**) was also identified and isolated from the sponge *Toxiclona toxius* [75]. Compounds **50**, **51**, **52** and **53** exhibited inhibitory activity of both DNA polymerizing functions of HIV-1 RT (RNA-dependent DNA polymerase RDDP and DNA-dependent DNA polymerase DDDP), but did not inhibit the RT-associated ribonuclease H activity. The hexaprenoid hydroquinones 50, 51, 52 and 53 inhibited RDDP function with IC_{50} values of 1.5, 3.3, 3.1 and 3.7 μM, and DDDP function with IC_{50} values of 6.6, 0.8, 2.7 and 8.2 μM, respectively [73]. Chemical investigation of *Suberea mollis*, a marine sponge collected from Hurghada, yielded the two brominated compounds subereaphenol B (**54**) and subereaphenol C (**55**) [76]. In a more recent study, subereaphenol C (**55**) was also isolated from the verongid sponge *Suberea* species, which was collected off Yanbu in Saudi Arabia [77]. Biological evaluation of compounds 54 and **55**, using the 2,2-diphenyl-1-picrylhydrazyl radical (DPPH) solution-based chemical assay, revealed a significant antioxidant activity (Figure 11). The high antioxidant activity could be due to the phenolic nature of the two compounds. None of compounds **54** and **55** exhibited any cytotoxicity against the human colon cancer cells, at a concentration of 10 μg/mL [76]. In another study, compound **55** showed antiproliferative activity against HeLa cells with IC_{50} value of 13.3 μM and no antimigratory activity against MDA-MB-321 cells (IC_{50} > 50 μM) [77]. The two brominated compounds moloka'iamine (**56**) and moloka'iakitamide (**57**) were isolated from the marine sponge *Pseudoceratina arabica*, which was collected from Sharm El-Sheikh [78] and Hurghada [79] in Egypt (Figure 11). Compounds 56 and 57 displayed significant parasympatholytic effects on isolated rabbit heart and jejunum, with no cytotoxicity against the HCT-116 cells, at a concentration of 10 μg/mL [78]. Moloka'iamine (**56**) was also isolated from the same marine sponge (*Pseudoceratina arabica*) collected from Anas Reef off Obhur in Saudi Arabia [80]. The dibrominated alkaloid ceratinine H (**58**) together with psammaplysin E (**59**) were isolated from the Red Sea marine Verongid sponge *Pseudoceratina arabica*, collected from Anas Reef off Obhur (Figure 11). The two alkaloids **58** and **59** were examined for their antimigratory activity against the highly metastatic human breast cancer cell line MDA-MB-231, and for their antiproliferative activity against HeLa cells. Ceratinine H (**58**) and psammaplysin E (**59**) exhibited antiproliferative activity against HeLa cells with IC_{50} values of 2.56 and 2.19 µM, respectively. Compound **59** displayed also high antimigratory activity against MDA-MB-231 cells with IC_{50} value of 0.31 µM [80]. Psammaplysin A (**60**) was isolated from the marine sponge *Pseudoceratina arabica*, which was obtained from both Anas Reef off Obhur at the Saudi coast [80] and Sharm El-Sheikh at the Egyptian coast [78]. In another report, psammaplysin A (**60**) and psammaplysin E (**59**) were isolated and identified from the verongid sponge *Aplysinella* species, harvested from Jizan, Saudi Arabia. Psammaplysin A (**60**) exhibited cytotoxic activity against MDA-MB-231, HeLa and HCT 116 cell lines with IC_{50} values of 3.9, 8.5 and 5.1 µM, respectively. While, psammaplysin E (**59**) showed a more potent cytotoxic activity against the same cell lines with IC_{50} values of 0.29, 2.1 and 3.7 µM, respectively [81].

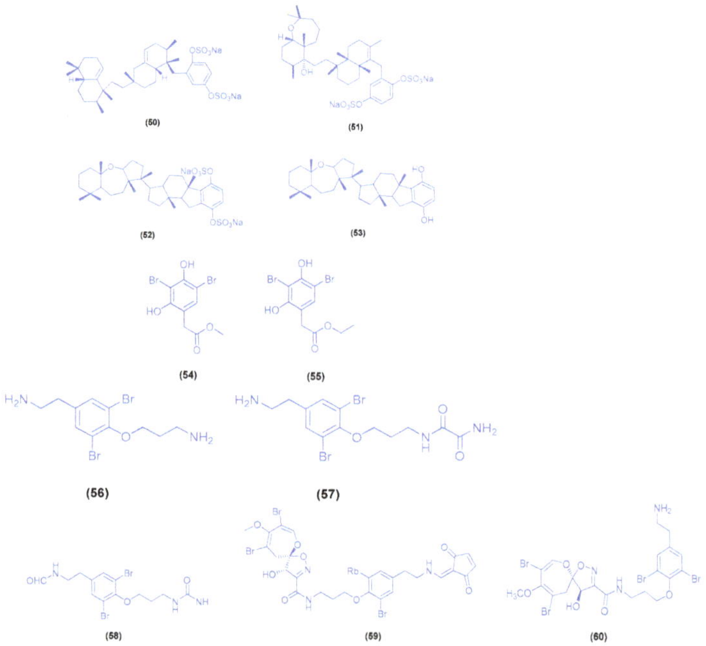

Figure 11. Chemical structures of toxiusol (**50**), shaagrockol C (**51**), toxicol A (**52**), toxicol B (**53**), subereaphenol B (**54**) and subereaphenol C (**55**), moloka'iamine (**56**), moloka'iakitamide (**57**), ceratinine H (**58**), psammaplysin E (**59**) and psammaplysin A (**60**).

The scalarane sesterterpene phyllospongin A (**61**), phyllospongin B (**62**), phyllospongin C (**63**), phyllospongin D (**64**), phyllospongin E (**65**), 12α-acetoxy20,24-dimethyl-25-norscalar-16-en-24-one (**66**), 12α-acetoxy-13β,18β-cyclobutan-20,24-dimethyl-24oxoscalar-16-en-25β-ol (**67**) and 12α-acetoxy-24,25-epoxy-24-hydroxy-20,24-dimethylscalarane (**68**) were isolated from the sponge *Phyllospongia lamellosa*, which was collected from Shaab Saad area northern Hurghada (Figure 12). Compounds **64**, **65** and **67** showed antibacterial activities against the Gram-positive pathogens *Staphylococcus aureus* ATCC25923 and *Bacillus subtilis* NCTC2116, with a range of MIC values from 1.7 to 2.5 µg/mL, and against the Gram-negative bacteria *Vibrio parahaemolyticus* NCTC10441 with MIC values of 6.8, 9.8 and 7 µg/mL, respectively. In addition, compounds **61–68** showed cytotoxic activities against three human cancer cell lines (HePG-2, MCF-7, and HCT-116), with a range of IC$_{50}$ values from 0.29 to 2.1 µg/mL [82].

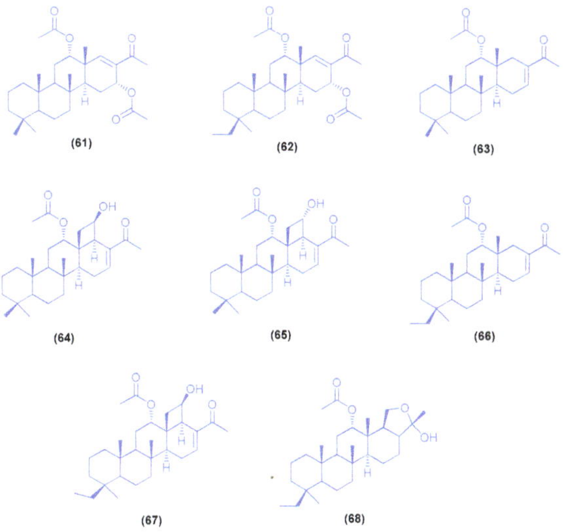

Figure 12. Chemical structures of phyllospongin A (**61**), phyllospongin B (**62**), phyllospongin C (**63**), phyllospongin D (**64**), phyllospongin E (**65**), compounds **66–68**.

The scalarane-type sesterterpenes 12β,20-αdihydroxy-16β-acetoxy-17-scalaren-19,20-olide (**69**), heteronemin (**70**) and 12-deacetyl-12,18-di-*epi*-scalaradial (**71**) were isolated and identified from the marine sponge *Hyrtios erectus*, collected from Sharm El-Sheikh in Egypt (Figure 13). Compounds **69–71** displayed remarkable antiproliferative activities against the breast adenocarcinoma MCF-7 cells (IC$_{50}$ = 12.7, 1.1 and 3.3 µM, respectively), the colorectal carcinoma HCT-116 cells (IC$_{50}$ = 3.5, 0.7 and 3.4 µM, respectively) and the hepatocellular carcinoma HepG2 cells (IC$_{50}$ = 9.6, 1.1 and 1.7 µM, respectively) [83]. The scalarane-type sesterterpenes sesterstatin 7 (**72**), 12-*epi*-24-deoxyscalarin (**73**) and 19-acetylsesterstatin 3 (**74**) were isolated from the sponge *Hyrtios erectus*, together with compounds **69–71** (Figure 13) [84]. Sesterstatin 7 (**72**) was also isolated from the sponge *Hyrtios erectus*, collected from Safaga in Egypt [85]. In another published study, compounds **72** and **73** were isolated from specimen of the marine sponge *Hyrtios erectus*, which was collected from Sharm El-Sheikh [83]. Compound **74** was also identified in another study and isolated from the sponge *Hyrtios erecta*, which was collected from El Quseir, 120 km south of Hurghada, Egypt [86]. Compounds **69**, **71–74** exhibited anti-tubercular activity against a nonvirulent *Mycobacterium tuberculosis* strain (ATCC 25177, H37Ra), with MIC values of 4.23, 5.05, 0.45, 1.12 and 4.39 µM, respectively. Sesterstatin 7 (**72**) exhibited also antibacterial activity against a strain of *Helicobacter pylori* (American Type Culture Collection, H.b., ATCC 700392), with MIC value of 4.39 µM. On the other hand, Compounds **69–71** showed significant cytotoxic activities against three human cancer cell lines (MCF-7, HCT-116 and HepG2) with a range IC$_{50}$ values ranging from 0.4 to 12.4 µM [84].

Chromatographic fractionation of the organic extract of the sponge *Theonella swinhoei*, collected from Hurghada, led to the isolation of the bicyclic glycopeptide theonellamide G (**75**) (Figure 14). Compound **75** exhibited significant antifungal activity against the wild and the amphotericin B-resistant strains of the fungal pathogen *Candida albicans*, with IC$_{50}$ values of 4.49 and 2 µM, respectively. It showed also cytotoxicity against HCT-16 cells with IC$_{50}$ value of 6 µM [87]. Bioassay-guided fractionation of the extract of the sponge *Suberea mollis*, collected from Hurghada, afforded the brominated alkaloid subereamolline A (**76**) (Figure 14) [76,79]. Compound **76** was a potent inhibitor of the migration and the invasion of MDA-MB-231 cells (a highly metastatic human breast cancer), with IC$_{50}$ value of 1.7 µM [79].

The steroidal glycoside eryloside A (**77**) was purified from the methanol extract of the marine sponge *Erylus lendenfeldi*, which was collected from the north of Hurghada (Figure 14). Compound **77** exhibited cytotoxicity against the yeast strain *Saccharomyces cerevisiae* with IC$_{50}$ value of 3.5 μM, and higher cytotoxicity against its mutant strain (deficient in double strand break repair), with IC$_{50}$ value of 0.8 μM [88]. The polychlorinated pyrrolidinone derivative dysidamide (**78**) was isolated from a sponge *Dysidea* sp. [89] and the sponge *Dysidea herbacea* [90]. In a later study, dysidamide (**78**) was isolated again from the dichloromethane extract of the marine sponge *Lamellodysidea herbacea*, which was collected from the Red Sea during the Ardoukoba expedition (Figure 14). Dysidamide (**78**) led to entire and rapid death of mesencephalic and cortical murine neurones at a dose of 0.8 μg/mL [91]. The two cytotoxic metabolites asmarine A (**79**) and asmarine B (**80**) were identified and isolated from the marine sponge *Raspailia* sp., collected from Dahlak archipelago, Eritrea (Figure 14) [92,93]. Compounds **79** and **80** were found to have cytotoxicity against P-388 murine leukemia cells, A-549 human lung carcinoma cells, HT-29 human colon carcinoma cells and MEL-28 human melanoma cells, with IC$_{50}$ range from 0.12 to 1.8 μM [92].

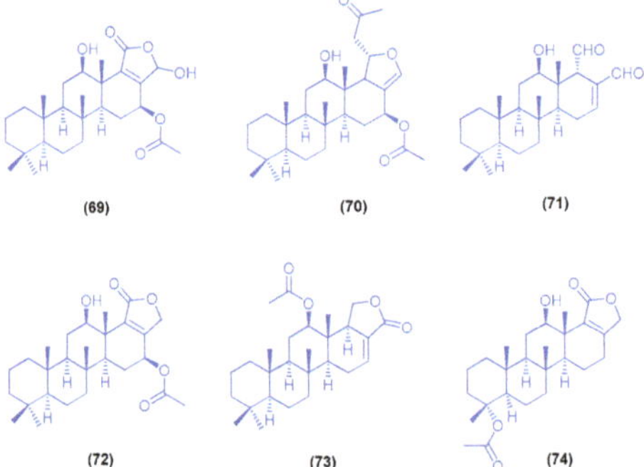

Figure 13. Chemical structures of compound **69**, heteronemin (**70**), compound **71**, sesterstatin 7 (**72**), 12-*epi*-24-deoxyscalarin (**73**) and 19 acetylsesterstatin 3 (**74**).

The two tripyridine alkaloids niphatoxin A (**81**) and niphatoxin B (**82**) were isolated from the sponge *Niphates* sp. (Figure 15). Compounds **81** and **82** showed cytotoxic activity against P-388 cells, with IC$_{50}$ value of 0.1 μg/mL [94]. The β-carboline alkaloid hyrtiomanzamine (**83**) was identified and isolated from the marine sponge *Hyrtios erecta* (Figure 15). Hyrtiomanzamine (**83**) showed immunosuppressive activity in the B lymphocytes reaction assay with EC$_{50}$ value of 2 μg/mL, with no cytotoxicity against KB cells [95]. The polyacetylene derivative petrosolic acid (**84**) was isolated from the sponge *Petrosia* sp. (Figure 15). Petrosolic acid (**84**) was tested for its inhibitory activity against various activities of HIV-1 RT (RNA-dependent DNA polymerase (RDDP), DNA-dependent DNA polymerase (DDDP) and RNase H functions), and showed inhibitory activity of RDDP and DDDP with IC$_{50}$ values of 1.2 and 5.9 μM, respectively [96]. The alkyl benzoate compound **85** and the oxysterol 3-β-hydroxycholest-5-en-7-one (**86**) were purified from the methylene chloride fraction of the marine sponge *Hyrtios erectus*, collected from the Saudi coast of Jeddah (Figure 15). Compounds **85** and **86** showed cytotoxic activities against MCF-7 cells, with IC$_{50}$ values of 2.4 and 3.8 μM, respectively. Compound **86** displayed also cytotoxicity against HepG 2 cells with IC$_{50}$ value of 1.3 μM [97]. Bioactivity-guided fractionation of the marine sponge *Callyspongia* aff. *Implexa*, collected from Safaga in Egypt, afforded the sterol compound gelliusterol E (**87**) (Figure 15). Compound **87** was tested against *Chlamydia trachomatis*, which is the leading cause of ocular and genital infections, and displayed antichlamydial activity in a dose-dependent manner with IC$_{50}$ value of 2.3 μM [98]. Two new bioactive brominated oxindole alkaloids **88** and **89** were recently isolated from the sponge *Callyspongia siphonella*, collected from Hurghada. The two metabolites 88 and 89 revealed antibacterial activity against *B. subtilis* with MIC value of 4 μg/mL and *S. aureus* (MIC = 8 and 16 μg/mL, respectively).

The two brominated oxindole alkaloids **88** and **89** showed also moderate in vitro antitrypanosomal activity and inhibited the biofilm formation of the Gram-negative bacterium *P. aeruginosa*. Both of **88** and **89** exhibited significant cytotoxicity against three cell lines, including HT-29, OVCAR-3 and MM.1S, with IC_{50} values ranging from 9 to 12.5 µM [59].

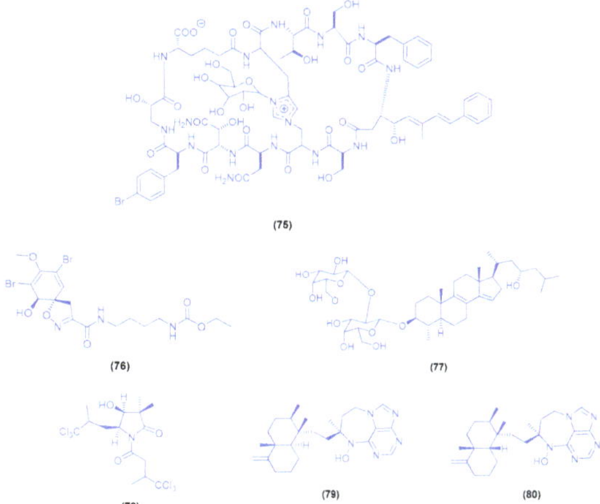

Figure 14. Chemical structures of theonellamide G (**75**), subereamolline A (**76**), eryloside A (**77**), dysidamide (**78**), asmarine A (**79**) and asmarine B (**80**).

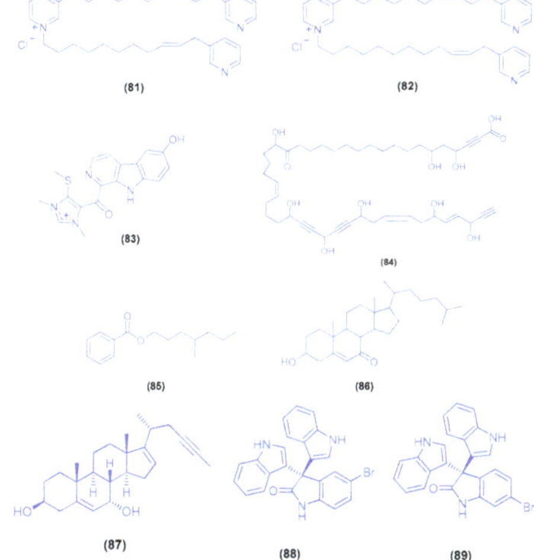

Figure 15. Chemical structures of niphatoxin A (**81**), niphatoxin B (**82**), hyrtiomanzamine (**83**), petrosolic acid (**84**), compounds **85–86**, gelliusterol E (**87**), compounds **88** and **89**.

3.2. Corals

The heptacyclic norcembranoid dimer singardin (**90**) and the sesquiterpene guaianediol (**91**) were obtained from the soft coral *Sinularia gardineri* (Pratt) (Alcyoniidae), collected near Hurghada in Egypt (Figure 16). Both compounds exhibited similar cytotoxic activity against murine leukemia (P-388) cells (IC_{50} = 1 µg/mL), human lung carcinoma (A-549) cells (IC_{50} = 2.5 µg/mL), human colon carcinoma (HT-29) cells (IC_{50} = 5 µg/mL) and human melanoma (MEL-28) cells (IC_{50} = 5 µg/mL) [99]. Bioassay-guided fractionation of the soft coral *Litophyton arboretum*, collected from Sharm

El-Sheikh, led to the isolation of 7β-acetoxy-24-methylcholesta-5-24(28)-diene-3,19-diol (**92**) and erythro-*N*-dodecanoyl-docosasphinga-(4*E*,8*E*)-dienine (**93**) (Figure 16). Compounds **92** and **93** displayed potent inhibitory activity against HIV-1 PR with IC_{50} values of 4.85 and 4.8 µM, respectively. Compound **92** exhibited also cytotoxicity against HeLa cells at CC_{50} value of 4.3 µM with selectivity index of 8.1, while compound **93** did not show cytotoxic activity [100]. Chemical investigations of the soft coral *Sarcophyton glaucum*, collected from Hurghada, afforded the peroxide diterpene 12(*S*)-hydroperoxylsarcoph-10-ene (**94**), together with 8-*epi*-sarcophinone (**95**) and *ent*-sarcophine (**96**) (Figure 16) [101]. Compounds **95** and **96** were also isolated from the soft coral *Sarcophyton trocheliophorum*, collected from the Hurghada [102]. Compounds **94–96** were inhibitors of the phase I enzyme cytochrome P450 1A with IC_{50} values of 2.7, 3.7 and 3.4 nM, respectively [101]. The polyhydroxylated sterol 97 together with the three ceramides **98–100** were isolated from the soft coral *Sinularia candidula*, collected from the Egyptian coast of Safaga (Figure 16). Compounds **97–100** exhibited antiviral activity against H5N1 virus and reduced the virus titer, at a concentration of 1 ng/mL, by 55.16%, 48.81%, 10.43% and 15.76%, respectively [103]. Fractionation of the organic extract of the soft coral *Sarcophyton glaucum*, collected from north of Jeddah in Saudi Arabia, yielded the two cembranoids sarcotrocheliol (**101**) and sarcotrocheliol acetate (**102**) (Figure 16). Sarcotrocheliol (**101**) and its acetate derivative 102 showed cytotoxic activity against MCF-7 cells with IC_{50} values of 2.4 and 3.2 µM, respectively [104].

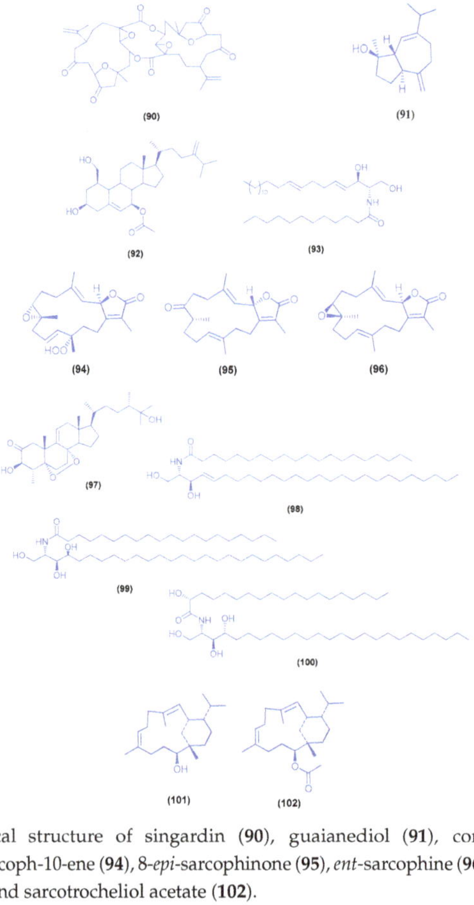

Figure 16. Chemical structure of singardin (**90**), guaianediol (**91**), compound **92** and **93**, 12(*S*)-hydroperoxylsarcoph-10-ene (**94**), 8-*epi*-sarcophinone (**95**), *ent*-sarcophine (**96**), compounds **97–100**, sarcotrocheliol (**101**) and sarcotrocheliol acetate (**102**).

3.3. Sea Hares

Oculiferane (**103**) and *epi*-obtusane (**104**) are two sesquiterpenes isolated from the acetone extract of the digestive gland of the sea hare *Aplysia oculifera*, which was collected from 40 km south of Safaga, Egypt (Figure 17). Compounds **103** and **104** showed in vitro cytotoxicity against several human cancer cell lines, including prostate carcinoma cells (PC-3), lung carcinoma (A549), human breast adenocarcinoma (MCF-7), hepatocellular carcinoma (HepG2) and colorectal carcinoma (HCT 116) cells, with a range of IC$_{50}$ values ranging from 0.96 to 5.9 μg/mL [105]. Isolation of dolastatin 16 (**4**) (Figure 5) from the sea hare *Dolabella auricularia* and its evaluation as anticancer agent were also reported. Compound **4** showed anticancer activity against several human cancer cell lines at a range of low micromolar concentrations (GI$_{50}$ = 0.0012–0.00096 μg/mL) [106].

Figure 17. Chemical structures of oculiferane (**103**) and *epi*-obtusane (**104**).

3.4. Tunicates

The diketopiperazine hydroxamate derivative etzionin (**105**) was isolated from an unidentified red tunicate, collected in the Northern part of the Gulf of Eilat in Israel (Figure 18). Etzionin (**105**) exhibited antifungal activity against the pathogenic yeast *Candida albicans* with MIC value of 3 μg/mL, in RPMI-1640 broth [107].

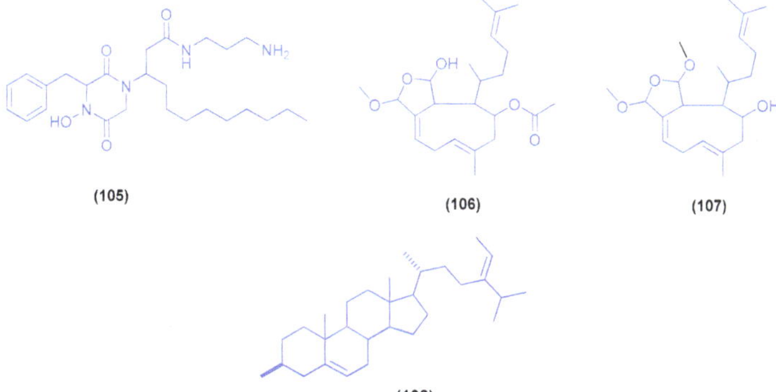

Figure 18. Chemical structures of etzionin (**105**), compound **106** and **107** and fucosterol (**108**).

4. Marine Natural Products Isolated from Marine Algae

Two cytotoxic xenicane diterpenes, 18,19-epoxyxenic-19-methoxy-18-hydroxy-4-acetoxy-6,9,13-triene (**106**) and 18,19-epoxyxenic-18,19-dimethoxy-4-hydroxy-6,9,13-triene (**107**), were purified from the methanol extract of the brown alga *Padina pavonia* (L.) Gaill., which was obtained from Hurghada, Egypt (Figure 18). The two cytotoxic xenicane diterpenes 106 and 107 showed significant in vitro cytotoxic activity against lung carcinoma (H460) cells at concentrations of 1–5 μg/mL [108]. The steroidal compound fucosterol (**108**) was isolated from the alcoholic extract of the brown alga *Dictyota dichotoma* (Hudson) Lamouroux, which was collected from Hurgada (Figure 18) [109]. Compound **108** was also isolated from the brown alga *Dictyota dichotoma* (Huds) Lamour, collected at Ras Abu-Bakr, 65 km north of Ras Gharib on Suez-Gulf, Egypt [110]. Compound **108** displayed potent cytotoxicity against mouse P 388 leukemia cells with IC$_{50}$ value of 0.6 μg/mL [109,111].

5. Sea Grasses

Bioactivity-guided fractionation of the extracts of the sea grass *Thallasodendron ciliatum*, collected from Egypt in Safaga [112] and Magawish near to Hurghada [113] led to the isolation of the diglyceride ester 109 and asebotin (**110**) (Figure 19). Anti-H5N1 virus activity of compounds **109** and **110** was measured using plaque inhibition assay in Madin–Darby canine kidney. Compounds **109** and **110** displayed reduction of virus titer by 67.26% and 53.81% inhibition at concentration of 1 ng/mL, respectively [112]. In a further study, the dihydrochalcone diglycoside thalassodendrone (**111**) together with asebotin (**110**) have been isolated from the ethyl acetate fraction of the sea grass *Thalassodendrin ciliatum* (Forsk.), which was obtained from Magawish city near Hurghada in Egypt (Figure 19). Asebotin (**110**) and thalassodendrone (**111**) exhibited antiviral activity against influenza A virus with IC_{50} values of 2 and 1.96 µg/mL, respectively, and with cytotoxic concentrations (CC_{50}) of 3.36 and 3.14 µg/mL, respectively [114].

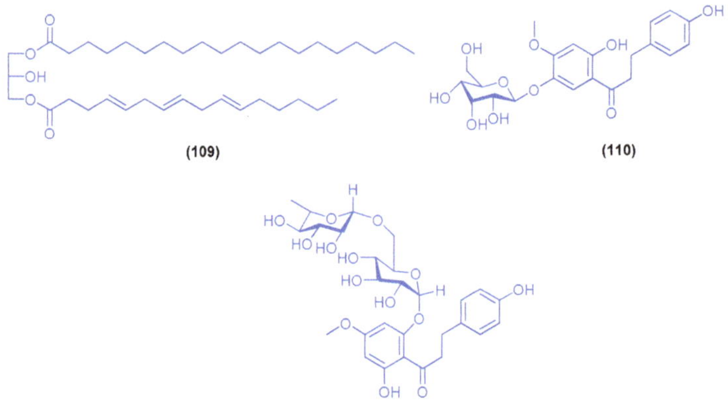

Figure 19. Chemical structures of compound **109**, asebotin (**110**) and thalassodendrone (**111**).

6. Genomic Potential of Red Sea Organisms

Metagenomic approach plays a crucial role in evaluating biodiversity from highly diverse and/or extreme habitats, e.g., marine environments, through direct access to genomes of inhabitant organisms. This has been potentiated through the integration of high-throughput DNA sequencing technologies and advanced-bioinformatic analyses, which highlighted the importance of metagenomics in bioprospecting novel bioactive metabolites from marine sources and their potential applications in biotechnology [115]. Metagenomic studies revealed that Red Sea harbors a unique microbial community with unique genes, enzymes and biosynthetic pathways, compared to other marine environments. This could be explained by the fact that Red Sea is regarded as an unusual ecosystem, characterized by its high salinity, high temperature, high ultra-violet radiation and low nutrients, as well as the existence of vibrant coral reefs and more than 25 hot brine pools [116]. In 2016, a group of researchers provided an overview of the previous metagenomic studies that were performed in the Red Sea [117]. Examination of the microbial community in the sediments of Kebrit Deep brine pool using 16srRNA gene PCR amplification, proved the presence of 16S sequences belonging to bacteria and Archaea unavailable in databases, suggesting the presence of novel microbial species harbored in Red Sea [118]. Further investigation of other brine pools in the Red Sea, gave accurate understanding about the microbial community composition and their unique metabolic pathways associated with the adaptation in such extreme environmental conditions [4,119].

The Red Sea coral *Stylophora pistillata* was found to live in symbiotic relationship with *Endozoicomonas* bacteria, that play a major role in the coral defense, health and survival. Studies demonstrated that diseased corals had lost their *Endozoicomonas* and has harbored opportunistic pathogens, including Alternaria and *Achromobacter*. Similarly, the symbiotic relationship between sponges, such as *Theonella swinhoei*, and their microbiomes had essentially contributed in the adaptation of the sponge to drastic environmental conditions and great tolerance to arsenite and arsenate [120]. Studies also revealed that the symbiotic association between the host and microbe can influence the microbial symbionts evolution. For instance, one of the unculturable sponge symbionts, the *cyanobaterium Candidatus Synechococcus spongiarum* associated with the Red Sea sponge *Carteriospongia foliascens*, has lost partial genes encoding proteins responsible for photosynthesis, production of polysaccharides, DNA repair mechanism and environmental stress adaptation, when compared to free-living cyanobacterial strains. The sponge tissues were collected from site RB4 (22°44′56″ N, 38°59′35″ E), located in the Rabigh Bay of Saudi Arabia along the Red Sea coast. Metagenomic DNA was sequenced on a 454 FLX platform utilizing Titanium chemistry, which produced a total of 315, 119 reads with a total length of 160.2 Mbp. Genome database were searched against the Kyoto Encyclopedia of Genes and Genomes (KEGG) database by using Basic Local Alignment Search Tool for Protein (BLASTP). Amino acid sequences were also searched against the GenBank NR data-base, and the output xml file was imported into MEGAN for taxonomic affiliation and SEED/Subsystems annotation. For the phylogenomic analysis, 31 proteins encoding phylogenetic markers were predicted from the SH4 draft genome and cyanobacterial genomes in the JGI database using AMPHORA. The sequences of each marker gene were aligned individually using ClustalW. The aligned sequences were concatenated, and a maximum-likelihood phylogenetic tree was constructed using PhyML [121].

The genome of Euryarchaea *Halorhabdus tiamatea*, obtained from the Shaban deep-sea hypersaline anoxic brine pool in the Red Sea, was sequenced by Werner et al. [122] in order to elucidate its niche adaptations. The study demonstrated that among the sequenced archaea, *H. tiamatea* enclosed the highest number of genes encoding glycoside hydrolases that could be potentially applied in industry. Glycosidase activity measurements suggested an adaptation towards recalcitrant algal and plant-derived hemicelluloses. Furthermore, *H. tiamatea* encoded proteins characteristic for thermophiles and light-dependent enzymes (e.g., bacteriorhodopsin), suggesting that *H. tiamatea* evolution was mostly not governed by a dark, cold, anoxic deep-sea habitat. The results supported that Halorhabdus species can occupy a distinct niche as polysaccharide degraders in hypersaline environments. *H. tiamatea* was sequenced on a 454 FLX Ti sequencer. Ribosomal RNA genes were identified via BLAST searches against public nucleotide databases and transfer RNA genes using TRNASCAN-SE v. 1.21. Selected CAZymes, from multi-functional CAZyme families, were subjected to in-depth phylogenetic. For each family, a set of experimentally characterized proteins was selected and aligned with their *H. tiamatea* homologues using MAFFT with iterative refinement and the Blosum62 matrix. Phylogenetic trees were computed from these alignments using PHYML.

Mohamed et al. [123] reported the largest number of Red Sea microbial genomes in a single study, with the aim to understand the microbial adaptation strategies to the extreme environmental conditions in Red Sea and allow the bioprospecting for novel thermo-and/or halo-philic enzymes. They have reported 136 microbial genomes assembled from 45 metagenomes, sampled form multiple depths (10–500 m), gradients in salinity, temperature, nutrients and oxygen, from 8 stations along the Red Sea. This represented great variation in environmental conditions as well as microbial diversity. The 136 retrieved genomes belonged to seven different phyla: *Thaumarchaeota, Euryarchaeota*, Actinobacteria, Cyanobacteria, *Bdellovibrionaeota*, Proteobacteria, and Marinimicrobia. Genomic DNA was extracted then sequenced on a HiSeq 2000 (Illumina, San Diego, CA, USA). Reads were quality checked and trimmed using PRINSEQ v0.20.4 generating read lengths of ~93 bp and a total of ~10 million reads per sample. Trimmed metagenome reads were individually assembled using IDBA-UD v1.1.1. The trimmed reads were mapped back to contigs using BWA v0.7.12 with the bwa-mem algorithm.

Ryu and coworkers [124] analyzed the hologenome data including genome, transcriptome, and metatranscriptome of two marine sponges; *Stylissa carteri* and *Xestospongia testudinaria* in order to investigate the host-symbiont interaction. Both species were collected at Fsar Reef (22.228408 N, 39.028187 E) on the Red Sea coast of Saudi Arabia, at a depth of 13–14 m. The paired-end and mate-pair sequencing was conducted using HiSeq2000 technology (Illumina), 454 and Ion proton sequencing were also performed using standard protocols. The Markov Cluster (MCL) Algorithm was utilized to cluster the Scavenger Receptor Cysteine-Rich (SRCR)-like domains together with the application of Blastp bit score as a similarity metric. The peptide sequences of the SRCR-like domain from each cluster were aligned using MAFFT v7. 123b and a custom Python script was used to compute the amino acid frequency at each aligned position. This is considered as a valuable and unique study because *S. carteri* is the first species in the order Halichondrida to have its genome sequenced whereas *X. testudinaria* is the first HMA sponge to have its genome sequenced. Results showed that *S. carteri*, when compared with *X. testudinaria*, has an expanded repertoire of immunological domains, in particular SRCR-like domains. Furthermore, an over-representation of potential symbiosis-related domains in *X. testudinari* was revealed through the metatranscriptome analyses.

The Red Sea sponge *Theonella swinhoei* was studied in order to investigate the uncultured majority of its symbionts. The sponge was collected from the Gulf of Aqaba, near Eilat. The DNA of *T. swinhoei* microbiome was sheared (200- to 400-bp size) and libraries were generated using the TruSeq DNA standard protocol and pooled for sequencing on one lane of the Illumina HiSeq 2000 platform. This study identified a complete *N*-acyl-homoserine lactone (AHL)-QS system (designated TswIR) and seven partial luxI homologues in the microbiome of *T. swinhoei*. The TswIR system was novel and shown to be associated with an alphaproteobacterium belonging to the order Rhodobacterales (termed Rhodobacterales bacterium TS309). When the gene tswI was expressed in Escherichia coli, it produced three AHLs, two of which were also identified in the studied sponge extract. The presence of some sponge-specific characteristics, such as ankyrin-like domains and tetratricopeptide repeats, the taxonomic affiliation of the 16S rRNA of Rhodobacterales bacterium TS309 to a sponge-coral specific clade, and its enrichment in sponge versus seawater and marine sediment samples, proved a likely symbiotic nature of this bacterium [125].

Recently, another group of researchers [126] focused on the isolation of *Pseudomonas* sp. associated with marine sponge *Hyrtios aff. erectus*, phylogenetic identification and molecular screening of their metabolic pathways polyketide synthases and nonribosomal peptides (PSK and NRPs). The sponge samples were obtained from the Red Sea at Hurghada city. The 16S rRNA gene sequencing of bacterial isolate demonstrated that active metabolic pathways genes in the *Pseudomonas* sp. was PSK II, which could be the reason for the antioxidant and cytotoxic bioactivity of the bacteria. The study, including the antioxidant, cytotoxicity and bioactive metabolic screening pathways, confirmed that this bacterial strain was shown to be an important source of natural antioxidant and cytotoxic metabolites against free radical and cancerous cell line.

7. Conclusions

The Red Sea is a rich and diverse ecosystem with 2000 km of coral reef extending along its coastline. It is inhabited by over 1000 invertebrate species, and 200 soft and hard corals. Due to this high biodiversity and limited research, the Red Sea is a promising underexplored habitat for the discovery of new bioactive marine natural products. Out of these, 677 natural products came from organisms isolated from the Red Sea, with almost 60% of these compounds coming from reports over the last decade (2011–2019) (Figures 20 and 21). This significant increase of reported compounds highlights the enormous potential of marine organisms, particularly from the Red Sea. Furthermore, the diversity of these chemical structures is quite remarkable. Majority of these compounds are represented by terpenes but also includes other classes namely alkaloids, sterols, peptides and other nitrogenous compounds, polyketides, fatty acids, macrolides, quinones, polyacetylene and flavonoids (Figure 22). It is interesting to note that from these different classes of natural products, majority of the compounds that were reported to exhibit the highest proportion of biological activities are terpenes followed by alkaloids. The presented marine natural products in this review (compounds **1–111**) exhibited a wide range of remarkable and potent biological activities, including for example, antioxidant, anticonvulsant, anticancer and anti-infective activities. However, biologically active marine natural products must be non-cytotoxic on normal cells, to be suitable for using as drugs. In the published studies, some of these biologically active marine natural products were tested for their cytotoxicity and found to be not toxic on normal cells. Therefore, these marine natural products could be promising compounds for developing new drugs after further investigations. Other biologically active marine natural products, which were not tested for their cytotoxicity, demand a testing for their cytotoxicity. Some of the illustrated cytotoxic marine natural products have PAINS characteristics (Pan-assay interference compounds), which could explain their cytotoxic activity. Among the organisms collected from the Red Sea, marine sponges remain to be the most widely sampled (Figure 3). This is not surprising since diverse species of marine invertebrates particularly sponges have been collected from the Red Sea and globally. Gram positive bacteria such as actinomycetes as well as fungi isolated from various marine organisms from the Red Sea have also been shown to be the prolific producers of these bioactive natural products (Figure 3). Furthermore, majority of the reported marine organisms have been collected from Egypt and Saudi Arabia coasts, accounting for 58% and 16% respectively. In contrast, less than 1% were isolated from Dijibouti, Jordan and Yemen (Figure 2). This could be attributed to the high biodiversity of organisms in Egyptian environment in comparison to other locations. Moreover, more research should be directed to explore other environments.

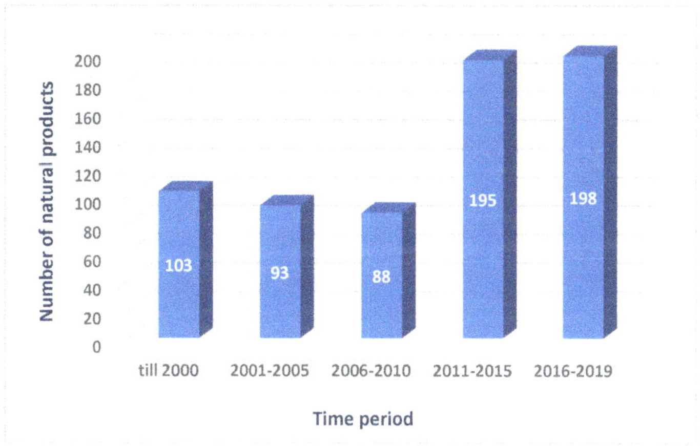

Figure 20. Numbers of natural products isolated from the Red Sea marine organisms.

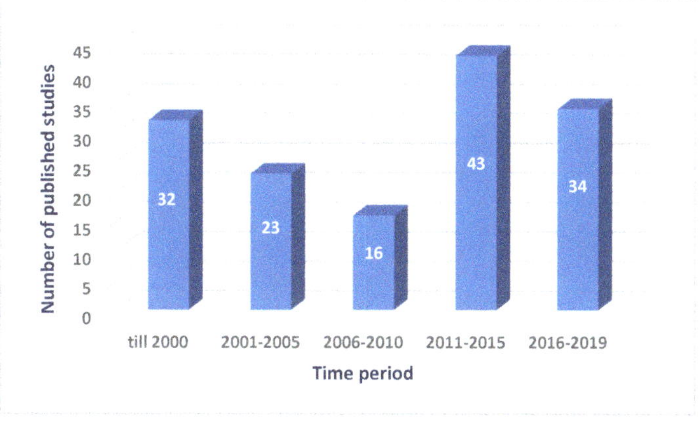

Figure 21. Numbers of published studies on natural products isolated from the Red Sea marine organisms.

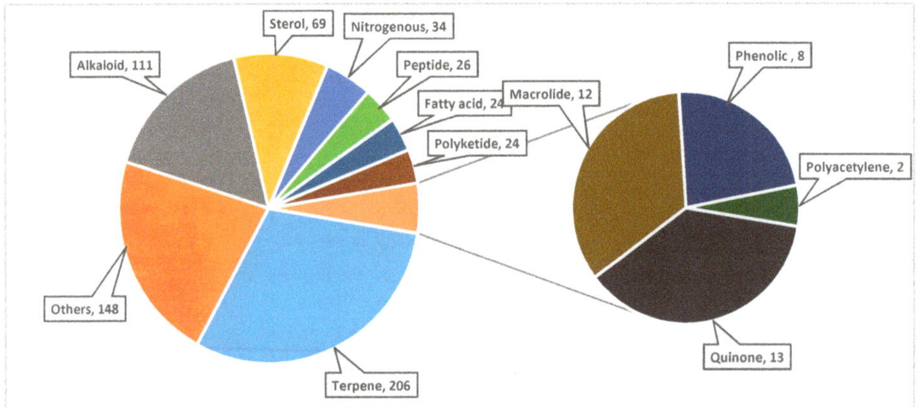

Figure 22. Chemical classes of natural products isolated from the Red Sea marine organisms.

Supplementary Materials: The following are available online at http://www.mdpi.com/1660-3397/18/9/457/s1. Table S1, Marine natural products of the Red Sea; Table S2: Locations of collection of the Red Sea marine organisms; Table S3: Taxonomy of marine organisms collected from the Red Sea.

Author Contributions: E.M.E.-H., M.A.-H., E.S.I., S.M.P.-E., H.H. and U.R.A. searched the literature; E.M.E.-H., M.A.-H., E.S.I., S.M.P.-E., H.H., M.F.A. and U.R.A. contributed to analyzing the collected data and writing the manuscript; E.M.E.-H., J.R.N., U.H. and U.R.A. contributed to reviewing and editing the final version of the manuscript. All authors have read and approved the final version of the manuscript.

Funding: This publication was supported by the Open Access Publication Fund of the University of Würzburg.

Acknowledgments: We would like to thank Minia University and Deraya University for supporting this work.

Conflicts of Interest: The authors declare there is no conflict of interest.

References

1. Bosworth, W.; Huchon, P.; McClay, K. The red sea and gulf of aden basins. *J. Afr. Earth Sci.* **2005**, *43*, 334–378. [CrossRef]
2. Raitsos, D.E.; Pradhan, Y.; Brewin, R.J.; Stenchikov, G.; Hoteit, I. Remote sensing the phytoplankton seasonal succession of the red sea. *PLoS ONE* **2013**, *8*, e64909. [CrossRef] [PubMed]
3. Berumen, M.L.; Hoey, A.S.; Bass, W.H.; Bouwmeester, J.; Catania, D.; Cochran, J.E.M.; Khalil, M.T.; Miyake, S.; Mughal, M.R.; Spaet, J.L.Y.; et al. The status of coral reef ecology research in the red sea. *Coral Reefs* **2013**, *32*, 737–748. [CrossRef]
4. Qian, P.Y.; Wang, Y.; Lee, O.O.; Lau, S.C.; Yang, J.; Lafi, F.F.; Al-Suwailem, A.; Wong, T.Y. Vertical stratification of microbial communities in the red sea revealed by 16s rdna pyrosequencing. *ISME J.* **2011**, *5*, 507–518. [CrossRef] [PubMed]
5. DiBattista, J.D.; Roberts, M.B.; Bouwmeester, J.; Bowen, B.W.; Coker, D.J.; Lozano-Cortés, D.F.; Howard Choat, J.; Gaither, M.R.; Hobbs, J.-P.A.; Khalil, M.T.; et al. A review of contemporary patterns of endemism for shallow water reef fauna in the red sea. *J. Biogeogr.* **2016**, *43*, 423–439. [CrossRef]
6. Ravara, A.; Carvalho, S. Nephtyidae (polychaeta, phyllodocida) from the red sea, with record of a new species. *J. Mar. Biol. Assoc. UK* **2017**, *97*, 843–856. [CrossRef]
7. Bowen, B.W.; Rocha, L.A.; Toonen, R.J.; Karl, S.A. The origins of tropical marine biodiversity. *Trends Ecol. Evol.* **2013**, *28*, 359–366. [CrossRef]
8. Gaither, M.R.; Bowen, B.W.; Bordenave, T.-R.; Rocha, L.A.; Newman, S.J.; Gomez, J.A.; van Herwerden, L.; Craig, M.T. Phylogeography of the reef fish cephalopholis argus(epinephelidae) indicates pleistocene isolation across the indo-pacific barrier with contemporary overlap in the coral triangle. *BMC Evol. Biol.* **2011**, *11*, 189. [CrossRef]

9. DiBattista, J.D.; Berumen, M.L.; Gaither, M.R.; Rocha, L.A.; Eble, J.A.; Choat, J.H.; Craig, M.T.; Skillings, D.J.; Bowen, B.W. After continents divide: Comparative phylogeography of reef fishes from the red sea and indian ocean. *J. Biogeogr.* **2013**, *40*, 1170–1181. [CrossRef]

10. Stark, J.S.; Kim, S.L.; Oliver, J.S. Anthropogenic disturbance and biodiversity of marine benthic communities in antarctica: A regional comparison. *PLoS ONE* **2014**, *9*, e98802. [CrossRef]

11. Mirto, S.; Gristina, M.; Sinopoli, M.; Maricchiolo, G.; Genovese, L.; Vizzini, S.; Mazzola, A. Meiofauna as an indicator for assessing the impact of fish farming at an exposed marine site. *Ecol. Indic.* **2012**, *18*, 468–476. [CrossRef]

12. Ziko, L.; Saqr, A.-H.A.; Ouf, A.; Gimpel, M.; Aziz, R.K.; Neubauer, P.; Siam, R. Antibacterial and anticancer activities of orphan biosynthetic gene clusters from atlantis ii red sea brine pool. *Microb. Cell Factories* **2019**, *18*, 56. [CrossRef] [PubMed]

13. Eisen, J.A. Environmental shotgun sequencing: Its potential and challenges for studying the hidden world of microbes. *PLoS Biol.* **2007**, *5*, e82. [CrossRef]

14. Trommer, G.; Siccha, M.; van der Meer, M.T.; Schouten, S.; Damsté, J.S.S.; Schulz, H.; Hemleben, C.; Kucera, M. Distribution of crenarchaeota tetraether membrane lipids in surface sediments from the red sea. *Org. Geochem.* **2009**, *40*, 724–731. [CrossRef]

15. Ellis, J.; Anlauf, H.; Kürten, S.; Lozano-Cortés, D.; Alsaffar, Z.; Cúrdia, J.; Jones, B.; Carvalho, S. Cross shelf benthic biodiversity patterns in the southern red sea. *Sci. Rep.* **2017**, *7*, 437. [CrossRef] [PubMed]

16. Hughes, D.J.; Lamont, P.A.; Levin, L.A.; Packer, M.; Feeley, K.; Gage, J.D. Macrofaunal communities and sediment structure across the pakistan margin oxygen minimum zone, north-east arabian sea. *Deep Sea Res. Part II Top. Stud. Oceanogr.* **2009**, *56*, 434–448. [CrossRef]

17. Cox, C.B.; Moore, P.D.; Ladle, R. *Biogeography: An Ecological and Evolutionary Approach*; John Wiley & Sons: Hoboken, NJ, USA, 2016.

18. DeLong, E.F.; Preston, C.M.; Mincer, T.; Rich, V.; Hallam, S.J.; Frigaard, N.-U.; Martinez, A.; Sullivan, M.B.; Edwards, R.; Brito, B.R. Community genomics among stratified microbial assemblages in the ocean's interior. *Science* **2006**, *311*, 496–503. [CrossRef] [PubMed]

19. Feebarani, J.; Joydas, T.; Damodaran, R.; Borja, A. Benthic quality assessment in a naturally-and human-stressed tropical estuary. *Ecol. Indic.* **2016**, *67*, 380–390. [CrossRef]

20. Silva, R.; Filho, J.R.; Souza, S.; Souza-Filho, P. Spatial and temporal changes in the structure of soft-bottom benthic communities in an amazon estuary (caeté estuary, brazil). *J. Coast. Res.* **2011**, 440–444.

21. Acker, J.; Leptoukh, G.; Shen, S.; Zhu, T.; Kempler, S. Remotely-sensed chlorophyll a observations of the northern red sea indicate seasonal variability and influence of coastal reefs. *J. Mar. Syst.* **2008**, *69*, 191–204. [CrossRef]

22. Sommer, B.; Harrison, P.L.; Beger, M.; Pandolfi, J.M. Trait-mediated environmental filtering drives assembly at biogeographic transition zones. *Ecology* **2014**, *95*, 1000–1009. [CrossRef] [PubMed]

23. Karakassis, I.; Eleftheriou, A. The continental shelf of crete: Structure of macrobenthic communities. *Mar. Ecol. Prog. Ser.* **1997**, *160*, 185–196. [CrossRef]

24. Vonk, J.A.; Stapel, J. Regeneration of nitrogen (15n) from seagrass litter in tropical indo-pacific meadows. *Mar. Ecol. Prog. Ser.* **2008**, *368*, 165–175. [CrossRef]

25. Sofianos, S.; Johns, W.E.; Murray, S. Heat and freshwater budgets in the red sea from direct observations at bab el mandeb. *Deep Sea Res. Part II Top. Stud. Oceanogr.* **2002**, *49*, 1323–1340. [CrossRef]

26. Pearman, J.K.; Irigoien, X.; Carvalho, S. Extracellular DNA amplicon sequencing reveals high levels of benthic eukaryotic diversity in the central red sea. *Mar. Genom.* **2016**, *26*, 29–39. [CrossRef] [PubMed]

27. Edwards, F.J. Climate and oceanography. *Red Sea* **1987**, *1*, 45–68.

28. Hasler, H.; Ott, J.A. Diving down the reefs? Intensive diving tourism threatens the reefs of the northern red sea. *Mar. Pollut. Bull.* **2008**, *56*, 1788–1794. [CrossRef]

29. Raitsos, D.; Hoteit, I.; Prihartato, P.; Chronis, T.; Triantafyllou, G.; Abualnaja, Y. Abrupt warming of the red sea. *Geophys. Res. Lett.* **2011**, *38*. [CrossRef]

30. Higgins, E.; Scheibling, R.E.; Desilets, K.M.; Metaxas, A. Benthic community succession on artificial and natural coral reefs in the northern gulf of aqaba, red sea. *PLoS ONE* **2019**, *14*, e0212842. [CrossRef]

31. El-Hossary, E.M.; Cheng, C.; Hamed, M.M.; El-Sayed Hamed, A.N.; Ohlsen, K.; Hentschel, U.; Abdelmohsen, U.R. Antifungal potential of marine natural products. *Eur. J. Med. Chem.* **2017**, *126*, 631–651. [CrossRef]

32. Abdelmohsen, U.R.; Balasubramanian, S.; Oelschlaeger, T.A.; Grkovic, T.; Pham, N.B.; Quinn, R.J.; Hentschel, U. Potential of marine natural products against drug-resistant fungal, viral, and parasitic infections. *Lancet Infect. Dis.* **2017**, *17*, e30–e41. [CrossRef]

33. Liu, M.; El-Hossary, E.M.; Oelschlaeger, T.A.; Donia, M.S.; Quinn, R.J.; Abdelmohsen, U.R. Potential of marine natural products against drug-resistant bacterial infections. *Lancet Infect. Dis.* **2019**, *19*, e237–e245. [CrossRef]

34. Shady, N.H.; El-Hossary, E.M.; Fouad, M.A.; Gulder, T.A.M.; Kamel, M.S.; Abdelmohsen, U.R. Bioactive natural products of marine sponges from the genus hyrtios. *Molecules* **2017**, *22*, 781. [CrossRef]

35. Pereira, F. Have marine natural product drug discovery efforts been productive and how can we improve their efficiency? *Expert Opin. Drug Discov.* **2019**, *14*, 717–722. [CrossRef] [PubMed]

36. Pereira, F.; Aires-de-Sousa, J. Computational methodologies in the exploration of marine natural product leads. *Mar. Drugs* **2018**, *16*, 236. [CrossRef]

37. Wiese, J.; Imhoff, J.F. Marine bacteria and fungi as promising source for new antibiotics. *Drug Dev. Res.* **2019**, *80*, 24–27. [CrossRef]

38. Newman, D.J.; Cragg, G.M. Drugs and drug candidates from marine sources: An assessment of the current "state of play". *Planta Med.* **2016**, *82*, 775–789. [CrossRef]

39. Gerwick, W.H.; Moore, B.S. Lessons from the past and charting the future of marine natural products drug discovery and chemical biology. *Chem. Biol.* **2012**, *19*, 85–98. [CrossRef]

40. El-Ezz, R.A.; Ibrahim, A.; Habib, E.; Wahba, A.; Kamel, H.; Afifi, M.; Hassanean, H.; Ahmed, S. Review of natural products from marine organisms in the red sea. *Int. J. Pharm. Sci. Res.* **2017**, *8*, 940–974.

41. Petitbois, J.G.; Casalme, L.O.; Lopez, J.A.V.; Alarif, W.M.; Abdel-Lateff, A.; Al-Lihaibi, S.S.; Yoshimura, E.; Nogata, Y.; Umezawa, T.; Matsuda, F.; et al. Serinolamides and lyngbyabellins from an *okeania* sp. Cyanobacterium collected from the red sea. *J. Nat. Prod.* **2017**, *80*, 2708–2715. [CrossRef]

42. Han, B.; McPhail, K.L.; Gross, H.; Goeger, D.E.; Mooberry, S.L.; Gerwick, W.H. Isolation and structure of five lyngbyabellin derivatives from a papua new guinea collection of the marine cyanobacterium lyngbya majuscula. *Tetrahedron* **2005**, *61*, 11723–11729. [CrossRef]

43. Tan, L.T.; Goh, B.P.; Tripathi, A.; Lim, M.G.; Dickinson, G.H.; Lee, S.S.; Teo, S.L. Natural antifoulants from the marine cyanobacterium lyngbya majuscula. *Biofouling* **2010**, *26*, 685–695. [CrossRef] [PubMed]

44. Lopez, J.A.; Al-Lihaibi, S.S.; Alarif, W.M.; Abdel-Lateff, A.; Nogata, Y.; Washio, K.; Morikawa, M.; Okino, T. Wewakazole B, a cytotoxic cyanobactin from the cyanobacterium moorea producens collected in the red sea. *J. Nat. Prod.* **2016**, *79*, 1213–1218. [CrossRef] [PubMed]

45. Thornburg, C.C.; Cowley, E.S.; Sikorska, J.; Shaala, L.A.; Ishmael, J.E.; Youssef, D.T.; McPhail, K.L. Apratoxin h and apratoxin a sulfoxide from the red sea cyanobacterium moorea producens. *J. Nat. Prod.* **2013**, *76*, 1781–1788. [CrossRef] [PubMed]

46. Thornburg, C.C.; Thimmaiah, M.; Shaala, L.A.; Hau, A.M.; Malmo, J.M.; Ishmael, J.E.; Youssef, D.T.; McPhail, K.L. Cyclic depsipeptides, grassypeptolides d and e and ibu-epidemethoxylyngbyastatin 3, from a red sea leptolyngbya cyanobacterium. *J. Nat. Prod.* **2011**, *74*, 1677–1685. [CrossRef] [PubMed]

47. Grkovic, T.; Abdelmohsen, U.R.; Othman, E.M.; Stopper, H.; Edrada-Ebel, R.; Hentschel, U.; Quinn, R.J. Two new antioxidant actinosporin analogues from the calcium alginate beads culture of sponge-associated *actinokineospora* sp. Strain eg49. *Bioorg. Med. Chem. Lett.* **2014**, *24*, 5089–5092. [CrossRef]

48. Tawfike, A.; Attia, E.Z.; Desoukey, S.Y.; Hajjar, D.; Makki, A.A.; Schupp, P.J.; Edrada-Ebel, R.; Abdelmohsen, U.R. New bioactive metabolites from the elicited marine sponge-derived bacterium actinokineospora *spheciospongiae* sp. Nov. *AMB Express* **2019**, *9*, 12. [CrossRef]

49. El-Gendy, M.M.; El-Bondkly, A.M. Production and genetic improvement of a novel antimycotic agent, saadamycin, against dermatophytes and other clinical fungi from endophytic *streptomyces* sp. Hedaya48. *J. Ind. Microbiol. Biotechnol.* **2010**, *37*, 831–841. [CrossRef]

50. Youssef, D.T.; Shaala, L.A.; Mohamed, G.A.; Ibrahim, S.R.; Banjar, Z.M.; Badr, J.M.; McPhail, K.L.; Risinger, A.L.; Mooberry, S.L. 2,3-seco-2,3-dioxo-lyngbyatoxin a from a red sea strain of the marine cyanobacterium moorea producens. *Nat. Prod. Res.* **2015**, *29*, 703–709. [CrossRef]

51. Shaala, L.A.; Youssef, D.T.; McPhail, K.L.; Elbandy, M. Malyngamide 4, a new lipopeptide from the red sea marine cyanobacterium moorea producens (formerly lyngbya majuscula). *Phytochem. Lett.* **2013**, *6*, 183–188. [CrossRef]

52. Shaala, L.A.; Youssef, D.T.; Badr, J.M.; Harakeh, S.M. Bioactive 2(1h)-pyrazinones and diketopiperazine alkaloids from a tunicate-derived actinomycete *Streptomyces* sp. *Molecules* **2016**, *21*, 1116. [CrossRef] [PubMed]

53. El-Gendy, M.M.; Hawas, U.W.; Jaspars, M. Novel bioactive metabolites from a marine derived bacterium *nocardia* sp. Alaa 2000. *J. Antibiot.* **2008**, *61*, 379–386. [CrossRef] [PubMed]

54. Elnaggar, M.S.; Ebada, S.S.; Ashour, M.L.; Ebrahim, W.; Müller, W.E.G.; Mándi, A.; Kurtán, T.; Singab, A.; Lin, W.; Liu, Z.; et al. Xanthones and sesquiterpene derivatives from a marine-derived fungus *Scopulariopsis* sp. *Tetrahedron* **2016**, *72*, 2411–2419. [CrossRef]

55. Lefranc, F.; Nuzzo, G.; Hamdy, N.A.; Fakhr, I.; Moreno, Y.B.L.; Van Goietsenoven, G.; Villani, G.; Mathieu, V.; van Soest, R.; Kiss, R.; et al. In vitro pharmacological and toxicological effects of norterpene peroxides isolated from the red sea sponge diacarnus erythraeanus on normal and cancer cells. *J. Nat. Prod.* **2013**, *76*, 1541–1547. [CrossRef]

56. Jain, S.; Laphookhieo, S.; Shi, Z.; Fu, L.W.; Akiyama, S.; Chen, Z.S.; Youssef, D.T.; van Soest, R.W.; El Sayed, K.A. Reversal of p-glycoprotein-mediated multidrug resistance by sipholane triterpenoids. *J. Nat. Prod.* **2007**, *70*, 928–931. [CrossRef]

57. Angawi, R.; Saqer, E.; Abdel-Lateff, A.; Badria, F.; Ayyad, S.-E. Cytotoxic neviotane triterpene-type from the red sea sponge *Siphonochalina siphonella*. *Pharmacogn. Mag.* **2014**, *10*, 334–341.

58. El-Beih, A.A.; El-Desoky, A.H.; Al-hammady, M.A.; Elshamy, A.I.; Hegazy, M.-E.F.; Kato, H.; Tsukamoto, S. New inhibitors of rankl-induced osteoclastogenesis from the marine sponge siphonochalina siphonella. *Fitoterapia* **2018**, *128*, 43–49. [CrossRef]

59. El-Hawary, S.S.; Sayed, A.M.; Mohammed, R.; Hassan, H.M.; Rateb, M.E.; Amin, E.; Mohammed, T.A.; El-Mesery, M.; Bin Muhsinah, A.; Alsayari, A.; et al. Bioactive brominated oxindole alkaloids from the red sea sponge callyspongia siphonella. *Mar. Drugs* **2019**, *17*, 465. [CrossRef]

60. Jain, S.; Abraham, I.; Carvalho, P.; Kuang, Y.H.; Shaala, L.A.; Youssef, D.T.; Avery, M.A.; Chen, Z.S.; El Sayed, K.A. Sipholane triterpenoids: Chemistry, reversal of abcb1/p-glycoprotein-mediated multidrug resistance, and pharmacophore modeling. *J. Nat. Prod.* **2009**, *72*, 1291–1298. [CrossRef]

61. Vilozny, B.; Amagata, T.; Mooberry, S.L.; Crews, P. A new dimension to the biosynthetic products isolated from the sponge negombata magnifica. *J. Nat. Prod.* **2004**, *67*, 1055–1057. [CrossRef]

62. El Sayed, K.A.; Youssef, D.T.; Marchetti, D. Bioactive natural and semisynthetic latrunculins. *J. Nat. Prod.* **2006**, *69*, 219–223. [CrossRef] [PubMed]

63. Ahmed, S.A.; Khalifa, S.I.; Hamann, M.T. Antiepileptic ceramides from the red sea sponge negombata corticata. *J. Nat. Prod.* **2008**, *71*, 513–515. [CrossRef] [PubMed]

64. Orabi, K.Y.; El Sayed, K.A.; Hamann, M.T.; Dunbar, D.C.; Al-Said, M.S.; Higa, T.; Kelly, M. Araguspongines k and l, new bioactive bis-1-oxaquinolizidine n-oxide alkaloids from red sea specimens of xestospongia exigua. *J. Nat. Prod.* **2002**, *65*, 1782–1785. [CrossRef] [PubMed]

65. El Sayed, K.A.; Hamann, M.T.; Hashish, N.E.; Shier, W.T.; Kelly, M.; Khan, A.A. Antimalarial, antiviral, and antitoxoplasmosis norsesterterpene peroxide acids from the red sea sponge diacarnus erythraeanus. *J. Nat. Prod.* **2001**, *64*, 522–524. [CrossRef]

66. Youssef, D.T.; Yoshida, W.Y.; Kelly, M.; Scheuer, P.J. Cytotoxic cyclic norterpene peroxides from a red sea sponge diacarnus e rythraenus. *J. Nat. Prod.* **2001**, *64*, 1332–1335. [CrossRef]

67. Youssef, D.T.A. Tasnemoxides a–c, new cytotoxic cyclic norsesterterpene peroxides from the red sea sponge diacarnus erythraeus. *J. Nat. Prod.* **2004**, *67*, 112–114. [CrossRef]

68. Youssef, D.T.; Mooberry, S.L. Hurghadolide a and swinholide i, potent actin-microfilament disrupters from the red sea sponge theonella swinhoei. *J. Nat. Prod.* **2006**, *69*, 154–157. [CrossRef]

69. Youssef, D.T.; van Soest, R.W.; Fusetani, N. Callyspongenols a-c, new cytotoxic c22-polyacetylenic alcohols from a red sea sponge, callyspongia species. *J. Nat. Prod.* **2003**, *66*, 679–681. [CrossRef]

70. Youssef, D.T.; van Soest, R.W.; Fusetani, N. Callyspongamide a, a new cytotoxic polyacetylenic amide from the red sea sponge callyspongia fistularis. *J. Nat. Prod.* **2003**, *66*, 861–862. [CrossRef]

71. Youssef, D.T. Hyrtioerectines a-c, cytotoxic alkaloids from the red sea sponge hyrtioserectus. *J. Nat. Prod.* **2005**, *68*, 1416–1419. [CrossRef]

72. Hirsch, S.; Rudi, A.; Kashman, Y.; Loya, Y. New avarone and avarol derivatives from the marine sponge dysidea cinerea. *J. Nat. Prod.* **1991**, *54*, 92–97. [CrossRef] [PubMed]

73. Loya, S.; Tal, R.; Hizi, A.; Issacs, S.; Kashman, Y.; Loya, Y. Hexaprenoid hydroquinones, novel inhibitors of the reverse transcriptase of human immunodeficiency virus type 1. *J. Nat. Prod.* **1993**, *56*, 2120–2125. [CrossRef]

74. Isaacs, S.; Hizi, A.; Kashman, Y. Toxicols ac and toxiusol-new bioactive hexaprenoid hydroquinones from toxiclona toxius. *Tetrahedron* **1993**, *49*, 4275–4282. [CrossRef]

75. Isaacs, S.; Kashman, Y. Shaagrockol B and C; two hexaprenylhydroquinone disulfates from the red sea sponge toxiclona toxius. *Tetrahedron Lett.* **1992**, *33*, 2227–2230. [CrossRef]

76. Abou-Shoer, M.I.; Shaala, L.A.; Youssef, D.T.; Badr, J.M.; Habib, A.A. Bioactive brominated metabolites from the red sea sponge suberea mollis. *J. Nat. Prod.* **2008**, *71*, 1464–1467. [CrossRef]

77. Shaala, L.; Youssef, D.; Badr, J.; Sulaiman, M.; Khedr, A. Bioactive secondary metabolites from the red sea marine verongid sponge suberea species. *Mar. Drugs* **2015**, *13*, 1621–1631. [CrossRef]

78. Badr, J.M.; Shaala, L.A.; Abou-Shoer, M.I.; Tawfik, M.K.; Habib, A.A. Bioactive brominated metabolites from the red sea sponge pseudoceratina arabica. *J. Nat. Prod.* **2008**, *71*, 1472–1474. [CrossRef]

79. Shaala, L.A.; Youssef, D.T.; Sulaiman, M.; Behery, F.A.; Foudah, A.I.; Sayed, K.A. Subereamolline a as a potent breast cancer migration, invasion and proliferation inhibitor and bioactive dibrominated alkaloids from the red sea sponge pseudoceratina arabica. *Mar. Drugs* **2012**, *10*, 2492–2508. [CrossRef]

80. Shaala, L.A.; Youssef, D.T.A.; Badr, J.M.; Sulaiman, M.; Khedr, A.; El Sayed, K.A. Bioactive alkaloids from the red sea marine verongid sponge pseudoceratina arabica. *Tetrahedron* **2015**, *71*, 7837–7841. [CrossRef]

81. Shaala, L.A.; Youssef, D.T.A. Cytotoxic psammaplysin analogues from the verongid red sea sponge aplysinella species. *Biomolecules* **2019**, *9*, 841. [CrossRef]

82. Hassan, M.H.A.; Rateb, M.E.; Hetta, M.; Abdelaziz, T.A.; Sleim, M.A.; Jaspars, M.; Mohammed, R. Scalarane sesterterpenes from the egyptian red sea sponge phyllospongia lamellosa. *Tetrahedron* **2015**, *71*, 577–583. [CrossRef]

83. Elhady, S.S.; Al-Abd, A.M.; El-Halawany, A.M.; Alahdal, A.M.; Hassanean, H.A.; Ahmed, S.A. Antiproliferative scalarane-based metabolites from the red sea sponge hyrtios erectus. *Mar. Drugs* **2016**, *14*, 130. [CrossRef] [PubMed]

84. Alahdal, A.M.; Asfour, H.Z.; Ahmed, S.A.; Noor, A.O.; Al-Abd, A.M.; Elfaky, M.A.; Elhady, S.S. Anti-helicobacter, antitubercular and cytotoxic activities of scalaranes from the red sea sponge hyrtios erectus. *Molecules* **2018**, *23*, 978. [CrossRef] [PubMed]

85. Youssef, D.T.; Shaala, L.A.; Emara, S. Antimycobacterial scalarane-based sesterterpenes from the red sea sponge hyrtios e recta. *J. Nat. Prod.* **2005**, *68*, 1782–1784. [CrossRef] [PubMed]

86. Youssef, D.T.; Yamaki, R.K.; Kelly, M.; Scheuer, P.J. Salmahyrtisol a, a novel cytotoxic sesterterpene from the red sea sponge hyrtios e recta. *J. Nat. Prod.* **2002**, *65*, 2–6. [CrossRef]

87. Youssef, D.T.; Shaala, L.A.; Mohamed, G.A.; Badr, J.M.; Bamanie, F.H.; Ibrahim, S.R. Theonellamide g, a potent antifungal and cytotoxic bicyclic glycopeptide from the red sea marine sponge theonella swinhoei. *Mar. Drugs* **2014**, *12*, 1911–1923. [CrossRef]

88. Sandler, J.S.; Forsburg, S.L.; Faulkner, D.J. Bioactive steroidal glycosides from the marine sponge erylus lendenfeldi. *Tetrahedron* **2005**, *61*, 1199–1206. [CrossRef]

89. Gebreyesusa, T.; Yosief, T.; Carmely, S.; Kashmanb, Y. Dysidamide, a novel hexachloro-metabolite from a red sea sponge *Dysidea* sp. *Tetrahedron Lett.* **1988**, *29*, 3863–3864. [CrossRef]

90. Carmely, S.; Gebreyesus, T.; Kashman, Y.; Skelton, B.; White, A.; Yosief, T. Dysidamide, a novel metabolite from a red sea sponge *Dysidea herbacea. Aust. J. Chem.* **1990**, *43*, 1881–1888. [CrossRef]

91. Sauleau, P.; Retailleau, P.; Vacelet, J.; Bourguet-Kondracki, M.-L. New polychlorinated pyrrolidinones from the red sea marine sponge lamellodysidea herbacea. *Tetrahedron* **2005**, *61*, 955–963. [CrossRef]

92. Yosief, T.; Rudi, A.; Stein, Z.; Goldberg, I.; Gravalos, G.M.D.; Schleyer, M.; Kashman, Y. Asmarines A–C; three novel cytotoxic metabolites from the marine sponge *Raspailia* sp. *Tetrahedron Lett.* **1998**, *39*, 3323–3326. [CrossRef]

93. Yosief, T.; Rudi, A.; Kashman, Y. Asmarines A–F, novel cytotoxic compounds from the marine sponge raspailia species. *J. Nat. Prod.* **2000**, *63*, 299–304. [CrossRef] [PubMed]

94. Talpir, R.; Rudi, A.; Ilan, M.; Kashman, Y. Niphatoxin a and b; two new ichthyo- and cytotoxic tripyridine alkaloids from a marine sponge. *Tetrahedron Lett.* **1992**, *33*, 3033–3034. [CrossRef]

95. Bourguet-Kondracki, M.L.; Martin, M.T.; Guyot, M. A new β-carboline alkaloid isolated from the marine sponge hyrtios erecta. *Tetrahedron Lett.* **1996**, *37*, 3457–3460. [CrossRef]

96. Isaacs, S.; Kashman, Y.; Loya, S.; Hizi, A.; Loya, Y. Petrosynol and petrosolic acid, two novel natural inhibitors of the reverse transcriptase of human immunodeficiency virus from *Petrosia* sp. *Tetrahedron* **1993**, *49*, 10435–10438. [CrossRef]

97. Hawas, U.W.; El-Kassem, L.T.A.; Abdelfattah, M.S.; Elmallah, M.I.Y.; Eid, M.A.G.; Monier, M.; Marimuthu, N. Cytotoxic activity of alkyl benzoate and fatty acids from the red sea sponge hyrtios erectus. *Nat. Prod. Res.* **2018**, *32*, 1369–1374. [CrossRef]

98. Abdelmohsen, U.R.; Cheng, C.; Reimer, A.; Kozjak-Pavlovic, V.; Ibrahim, A.K.; Rudel, T.; Hentschel, U.; Edrada-Ebel, R.; Ahmed, S.A. Antichlamydial sterol from the red sea sponge callyspongia aff. Implexa. *Planta Med.* **2015**, *81*, 382–387. [CrossRef]

99. El Sayed, K.A.; Hamann, M.T. A new norcembranoid dimer from the red sea soft coral sinularia gardineri. *J. Nat. Prod.* **1996**, *59*, 687–689. [CrossRef]

100. Ellithey, M.S.; Lall, N.; Hussein, A.A.; Meyer, D. Cytotoxic, cytostatic and hiv-1 pr inhibitory activities of the soft coral litophyton arboreum. *Mar. Drugs* **2013**, *11*, 4917–4936. [CrossRef]

101. Hegazy, M.E.; Gamal Eldeen, A.M.; Shahat, A.A.; Abdel-Latif, F.F.; Mohamed, T.A.; Whittlesey, B.R.; Pare, P.W. Bioactive hydroperoxyl cembranoids from the red sea soft coral sarcophyton glaucum. *Mar. Drugs* **2012**, *10*, 209–222. [CrossRef]

102. Hegazy, M.-E.F.; Mohamed, T.A.; Abdel-Latif, F.F.; Alsaid, M.S.; Shahat, A.A.; Pare, P.W. Trochelioid a and b, new cembranoid diterpenes from the red sea soft coral sarcophyton trocheliophorum. *Phytochem. Lett.* **2013**, *6*, 383–386. [CrossRef]

103. Ahmed, S.; Ibrahim, A.; Arafa, A.S. Anti-h5n1 virus metabolites from the red sea soft coral, sinularia candidula. *Tetrahedron Lett.* **2013**, *54*, 2377–2381. [CrossRef]

104. Abdel-Lateff, A.; Alarif, W.M.; Ayyad, S.E.; Al-Lihaibi, S.S.; Basaif, S.A. New cytotoxic isoprenoid derivatives from the red sea soft coral sarcophyton glaucum. *Nat. Prod. Res.* **2015**, *29*, 24–30. [CrossRef]

105. Hegazy, M.-F.F.; Moustfa, A.Y.; Mohamed, A.E.-H.H.; Alhammady, M.A.; Elbehairi, S.E.I.; Ohta, S.; Paré, P.W. New cytotoxic halogenated sesquiterpenes from the egyptian sea hare, aplysia oculifera. *Tetrahedron Lett.* **2014**, *55*, 1711–1714. [CrossRef]

106. Pettit, G.R.; Xu, J.P.; Hogan, F.; Williams, M.D.; Doubek, D.L.; Schmidt, J.M.; Cerny, R.L.; Boyd, M.R. Isolation and structure of the human cancer cell growth inhibitory cyclodepsipeptide dolastatin 16. *J. Nat. Prod.* **1997**, *60*, 752–754. [CrossRef] [PubMed]

107. Hirsch, S.; Miroz, A.; McCarthy, P.; Kashman, Y. Etzionin, a new antifungal metabolite from a red sea tunicate. *Tetrahedron Lett.* **1989**, *30*, 4291–4294. [CrossRef]

108. Awad, N.E.; Selim, M.A.; Metawe, H.M.; Matloub, A.A. Cytotoxic xenicane diterpenes from the brown alga padina pavonia (l.) gaill. *Phytother. Res.* **2008**, *22*, 1610–1613. [CrossRef]

109. Gedara, S.R.; Abdel-Halim, O.B.; el-Sharkawy, S.H.; Salama, O.M.; Shier, T.W.; Halim, A.F. Cytotoxic hydroazulene diterpenes from the brown alga dictyota dichotoma. *Z. Nat. C* **2003**, *58*, 17–22. [CrossRef]

110. Abou-El-Wafa, G.S.; Shaaban, M.; Shaaban, K.A.; El-Naggar, M.E.; Maier, A.; Fiebig, H.H.; Laatsch, H. Pachydictyols b and c: New diterpenes from dictyota dichotoma hudson. *Mar. Drugs* **2013**, *11*, 3109–3123. [CrossRef]

111. Sheu, J.H.; Wang, G.H.; Sung, P.J.; Chiu, Y.H.; Duh, C.Y. Cytotoxic sterols from the formosan brown alga turbinaria ornata. *Planta Med.* **1997**, *63*, 571–572. [CrossRef]

112. Ibrahim, A.K.; Youssef, A.I.; Arafa, A.S.; Foad, R.; Radwan, M.M.; Ross, S.; Hassanean, H.A.; Ahmed, S.A. Anti-h5n1 virus new diglyceride ester from the red sea grass thallasodendron ciliatum. *Nat. Prod. Res.* **2013**, *27*, 1625–1632. [CrossRef]

113. Hamdy, A.H.; Mettwally, W.S.; El Fotouh, M.A.; Rodriguez, B.; El-Dewany, A.I.; El-Toumy, S.A.; Hussein, A.A. Bioactive phenolic compounds from the egyptian red sea seagrass thalassodendron ciliatum. *Z. Nat. C* **2012**, *67*, 291–296.

114. Mohammed, M.M.; Hamdy, A.H.; El-Fiky, N.M.; Mettwally, W.S.; El-Beih, A.A.; Kobayashi, N. Anti-influenza a virus activity of a new dihydrochalcone diglycoside isolated from the egyptian seagrass thalassodendron ciliatum (forsk.) den hartog. *Nat. Prod. Res.* **2014**, *28*, 377–382. [CrossRef] [PubMed]

115. Barone, R.; De Santi, C.; Palma Esposito, F.; Tedesco, P.; Galati, F.; Visone, M.; di Scala, A.; de Pascale, D. Marine metagenomics, a valuable tool for enzymes and bioactive compounds discovery. *Front. Mar. Sci.* **2014**, *1*. [CrossRef]

116. Gurvich, E.G. *Metalliferous Sediments of the World Ocean: Fundamental Theory of Deep-Sea Hydrothermal Sedimentation*; Springer: Berlin/Heidelberg, Germany, 2006.
117. Behzad, H.; Ibarra, M.A.; Mineta, K.; Gojobori, T. Metagenomic studies of the red sea. *Gene* **2016**, *576*, 717–723. [CrossRef] [PubMed]
118. Eder, W.; Ludwig, W.; Huber, R. Novel 16s rrna gene sequences retrieved from highly saline brine sediments of kebrit deep, red sea. *Arch. Microbiol.* **1999**, *172*, 213–218. [CrossRef] [PubMed]
119. Thompson, L.R.; Field, C.; Romanuk, T.; Ngugi, D.; Siam, R.; El Dorry, H.; Stingl, U. Patterns of ecological specialization among microbial populations in the red sea and diverse oligotrophic marine environments. *Ecol. Evol.* **2013**, *3*, 1780–1797. [CrossRef]
120. Bayer, T.; Neave, M.J.; Alsheikh-Hussain, A.; Aranda, M.; Yum, L.K.; Mincer, T.; Hughen, K.; Apprill, A.; Voolstra, C.R. The microbiome of the red sea coral stylophora pistillata is dominated by tissue-associated endozoicomonas bacteria. *Appl. Environ. Microbiol.* **2013**, *79*, 4759–4762. [CrossRef] [PubMed]
121. Gao, Z.M.; Wang, Y.; Tian, R.M.; Wong, Y.H.; Batang, Z.B.; Al-Suwailem, A.M.; Bajic, V.B.; Qian, P.Y. Symbiotic adaptation drives genome streamlining of the cyanobacterial sponge symbiont "candidatus synechococcus spongiarum". *mBio* **2014**, *5*, e00079-14. [CrossRef] [PubMed]
122. Werner, J.; Ferrer, M.; Michel, G.; Mann, A.J.; Huang, S.; Juarez, S.; Ciordia, S.; Albar, J.P.; Alcaide, M.; la Cono, V.; et al. Halorhabdus tiamatea: Proteogenomics and glycosidase activity measurements identify the first cultivated euryarchaeon from a deep-sea anoxic brine lake as potential polysaccharide degrader. *Environ. Microbiol.* **2014**, *16*, 2525–2537. [CrossRef] [PubMed]
123. Haroon, M.F.; Thompson, L.R.; Parks, D.H.; Hugenholtz, P.; Stingl, U. A catalogue of 136 microbial draft genomes from red sea metagenomes. *Sci. Data* **2016**, *3*, 160050. [CrossRef]
124. Ryu, T.; Seridi, L.; Moitinho-Silva, L.; Oates, M.; Liew, Y.J.; Mavromatis, C.; Wang, X.; Haywood, A.; Lafi, F.F.; Kupresanin, M.; et al. Hologenome analysis of two marine sponges with different microbiomes. *BMC Genom.* **2016**, *17*, 158. [CrossRef] [PubMed]
125. Britstein, M.; Devescovi, G.; Handley, K.M.; Malik, A.; Haber, M.; Saurav, K.; Teta, R.; Costantino, V.; Burgsdorf, I.; Gilbert, J.A.; et al. A new n-acyl homoserine lactone synthase in an uncultured symbiont of the red sea sponge theonella swinhoei. *Appl. Environ. Microbiol.* **2016**, *82*, 1274–1285. [CrossRef] [PubMed]
126. El-Moneam, N.; El-Assar, S.; Shreadah, M.; Adam, A. Isolation, identification and molecular screening of *pseudomonas* sp. Metabolic pathways nrps and pks associated with the red sea sponge, hyrtios aff. Erectus, egypt. *J. Pure Appl. Microbiol.* **2017**, *11*, 1299–1311. [CrossRef]

 *marine drugs*

Article

Azaphilones from the Red Sea Fungus *Aspergillus falconensis*

Dina H. El-Kashef [1,2], Fadia S. Youssef [1,3], Rudolf Hartmann [4], Tim-Oliver Knedel [5], Christoph Janiak [5], Wenhan Lin [6], Irene Reimche [7], Nicole Teusch [7], Zhen Liu [1,*] and Peter Proksch [1,8,*]

1   Institute of Pharmaceutical Biology and Biotechnology, Heinrich-Heine-University Duesseldorf, 40225 Duesseldorf, Germany; dina.elkashef@mu.edu.eg (D.H.E.-K.); fadiayoussef@pharma.asu.edu.eg (F.S.Y.)

2   Department of Pharmacognosy, Faculty of Pharmacy, Minia University, 61519 Minia, Egypt

3   Department of Pharmacognosy, Faculty of Pharmacy, Ain Shams University, Abbassia, 11566 Cairo, Egypt

4   Institute of Complex Systems: Strukturbiochemie, Forschungszentrum Jülich GmbH, ICS-6, 52425 Jülich, Germany; r.hartmann@fz-juelich.de

5   Institut für Anorganische Chemie und Strukturchemie, Heinrich-Heine-Universität Düsseldorf, 40225 Düsseldorf, Germany; tim-oliver.knedel@hhu.de (T.-O.K.); janiak@uni-duesseldorf.de (C.J.)

6   State Key Laboratory of Natural and Biomimetic Drugs, Peking University, Beijing 100191, China; whlin@bjmu.edu.cn

7   Department of Biomedical Sciences, Institute of Health Research and Education, University of Osnabrück, 49074 Osnabrück, Germany; irene.reimche@uni-osnabrueck.de (I.R.); nicole.teusch@uni-osnabrueck.de (N.T.)

8   Hubei Key Laboratory of Natural Products Research and Development, College of Biological and Pharmaceutical Sciences, China Three Gorges University, Yichang 443002, China

*   Correspondence: zhenfeizi0@sina.com (Z.L.); proksch@uni-duesseldorf.de (P.P.); Tel.: +49-211-81-14163 (Z.L. & P.P.)

Received: 4 March 2020; Accepted: 5 April 2020; Published: 10 April 2020

Abstract: The marine-derived fungus *Aspergillus falconensis*, isolated from sediment collected from the Canyon at Dahab, Red Sea, yielded two new chlorinated azaphilones, falconensins O and P (**1** and **2**) in addition to four known azaphilone derivatives (**3−6**) following fermentation of the fungus on solid rice medium containing 3.5% NaCl. Replacing NaCl with 3.5% NaBr induced accumulation of three additional new azaphilones, falconensins Q−S (**7−9**) including two brominated derivatives (**7** and **8**) together with three known analogues (**10−12**). The structures of the new compounds were elucidated by 1D and 2D NMR spectroscopy and HRESIMS data as well as by comparison with the literature. The absolute configuration of the azaphilone derivatives was established based on single-crystal X-ray diffraction analysis of **5**, comparison of NMR data and optical rotations as well as on biogenetic considerations. Compounds **1**, **3−9**, and **11** showed NF-κB inhibitory activity against the triple negative breast cancer cell line MDA-MB-231 with IC_{50} values ranging from 11.9 to 72.0 μM.

Keywords: *Aspergillus falconensis*; OSMAC; azaphilones; X-ray diffraction; NF-κB inhibition

---

1. Introduction

In the last two decades, marine-derived fungi have gained considerable attention for drug discovery due to their ability to produce a vast diversity of bioactive secondary metabolites [1]. Up until today, hundreds of secondary metabolites have been characterized from marine-derived fungi exhibiting promising biological and pharmacological properties [2,3]. In particular, the diketopiperazine alkaloid halimide, obtained from a marine-derived fungus *Aspergillus* sp., was the lead structure for the putative anticancer drug plinabulin, which has entered Phase III of clinical trials against non-small cell lung cancer [4–6]. Interestingly, genomic sequencing revealed that under conventional culture conditions, many biosynthetic fungal gene clusters remain silent and transcriptionally suppressed [7,8]. Changing the cultivation conditions of fungi may activate silent biosynthetic gene clusters and eventually lead to either upregulation of constitutively present compounds or accumulation of new natural products [9]. The OSMAC (One Strain MAny Compounds) approach, first described by Zeeck et al. [10], represents one of the strategies which triggers diversification of the metabolic profile of fungi by altering the cultivation conditions. Fungi of the genus *Aspergillus* are rich sources of numerous bioactive secondary metabolites [11–13]. Consequently, as a part of our ongoing research on marine-derived fungi [14–16], we have investigated the fungus *Aspergillus falconensis* (formerly known as *Emericella falconensis* [17]) that was isolated from sea sediment collected at a depth of 25 m from the Canyon at Dahab, Red Sea, Egypt. The previously isolated soil-derived fungus, *Emericella falconesis*, is known as a producer of anti-inflammatory azaphilone derivatives [18–21]. Herein, we report the isolation, structure elucidation, and bioactivity of azaphilones obtained from *A. falconensis* following fermentation of the fungus on solid rice medium that contained either 3.5% NaCl or 3.5% NaBr. Following this approach, we were able to obtain two new chlorinated azaphilones (**1** and **2**) together with four known azaphilone derivatives (**3–6**) when the fungus was cultivated in the presence of NaCl and three additional new azaphilone derivatives (**7–9**) including two brominated analogues (**7** and **8**) in addition to three known derivatives (**10–12**) in the presence of NaBr. Compounds **1**, **3–9**, and **11** were examined with regard to their nuclear factor kappa B (NF-κB) inhibitory activity in the triple negative breast cancer (TNBC) cell line MDA-MB-231. In TNBC, constitutive activation of the proinflammatory NF-κB is associated with tumor aggressiveness [22]. All tested compounds revealed inhibition of NF-κB signaling with IC_{50} values at two-digit micromolar concentrations.

2. Results and Discussion

After fermentation of *A. falconensis* on solid rice medium containing 3.5% NaCl (similar to the salinity of sea water), chromatographic separation of the EtOAc extract of the fungus yielded two new chlorinated azaphilone derivatives (**1–2**). In addition, four known azaphilones (**3–6**) were identified including falconensins A (**3**) [18], M (**4**) [23], N (**5**) [23], and H (**6**) [19] by comparison of their spectroscopic data with the literature.

Compound **1** was isolated as a yellow oil. Its HRESIMS analysis showed an isotope pattern characteristic for two chlorine atoms in the molecule at *m/z* 511/513/515 (9:6:1), corresponding to the molecular formula $C_{24}H_{24}Cl_2O_8$ with 12 degrees of unsaturation. The UV pattern and NMR data of **1** (Tables 1 and 2) were similar to those of the co-isolated known falconensin M (**4**), which was previously reported from *Emericella falconensis* [23], suggesting **1** to be an azaphilone derivative [24]. However, the ^1H NMR spectrum of **1** displayed the signal of an additional methyl group at δ_H 2.07 (s) when compared to **4**. The Heteronuclear Multiple Bond Correlation (HMBC) correlations from the protons of this additional methyl and H-8 to a carbonyl carbon at δ_C 170.0 together with the obvious deshielded chemical shift of H-8 (δ_H 6.11) indicated the replacement of the hydroxy group by an acetoxy group at C-8 in **1**. Detailed analysis of the 2D NMR spectra of **1** indicated that the remaining substructure of **1** was identical to that of **4**. The large values of $^3J_{\text{H-8,H-8a}}$ (10.7 Hz) and $^3J_{\text{H-1b,H-8a}}$ (13.0 Hz) and the small value of $^3J_{\text{H-1a,H-8a}}$ (5.0 Hz) obtained from the ^1H NMR spectrum of **1**, indicated *trans*-diaxial orientation between H-8a (δ_H 2.98) and H-8 and between H-8a and H-1b (δ_H 3.96) and *cis*-orientation between H-8a and H-1a (δ_H 4.34). The NOE correlations from H-8a to H-1a and Me-9 (δ_H 1.56) as well as between H-8 and H-1b confirmed that H-1a, H-8a and Me-9 were on the same side of the ring (α-oriented) whereas H-1b and H-8 were on the opposite side (β-oriented). Thus, compound **1** was elucidated as 8-*O*-acetyl analogue of **4**, representing a new falconensin derivative, for which the trivial name falconensin O is proposed. Incubation of putative precursors of **1** such as **4** in EtOAc over three days failed to generate the corresponding acetates, thus ruling out the possibility that the acetylated azaphilones isolated in this study are artefacts generated during chromatographic workup.

Table 1. ^{13}C NMR data of compounds **1**, **2**, **7**, **8**, and **9** in CDCl$_3$.

No.	1 [a,c]	2 [a,c]	7 [b]	8 [a,c]	9 [b]
1	67.8, CH$_2$	68.2, CH$_2$	68.5, CH$_2$	68.4, CH$_2$	68.0, CH$_2$
3	160.2, C	168.2, C	160.5, C	160.2, C	160.2, C
4	102.4, CH	100.9, CH	102.8, CH	102.6 CH	102.8, CH
4a	149.4, C	149.2, C	150.3, C	149.9, C	148.9, C
5	116.6, CH	115.7, CH	116.6, CH	116.6, CH	117.1, CH
6	192.3, C	192.7, C	193.8, C	193.4, C	193.3, C
7	83.4, C	83.4, C	85.6, C	85.4, C	82.5, C
8	70.1, CH	70.4, CH	69.8, CH	69.7, CH	70.7, CH
8a	38.1, CH	38.0, CH	37.8, CH	37.5, CH	38.2, CH
9	17.9, CH$_3$	18.2, CH$_3$	16.8, CH$_3$	16.6, CH$_3$	18.2, CH$_3$
10	124.9, CH	36.4, CH$_2$	125.4, CH	125.3, CH	125.2, CH
11	134.1, CH	20.0, CH$_2$	133.9, CH	133.6, CH	133.9, CH
12	18.2, CH$_3$	13.6, CH$_3$	18.4, CH$_3$	18.2, CH$_3$	18.4, CH$_3$
1'	164.3, C	164.4, C	165.3, C	164.9, C	166.2, C
2'	122.2, C	122.3, C	117.4, C	117.2, C	115.5, C
3'	152.8, C	152.9, C	156.2, C	157.3, C	159.0, C
4'	112.9, C	112.8, C	97.1, CH	93.8, CH	96.9, CH
5'	149.2, C	149.8, C	154.0, C	155.8, C	157.6, C
6'	117.0, C	117.1, C	105.3, C	106.6, C	109.2, CH
7'	134.3, C	134.5, C	137.4, C	138.1, C	140.0, C
8'	16.9, CH$_3$	17.1, CH$_3$	20.2, CH$_3$	20.1, CH$_3$	19.6, CH$_3$
8-OAc	170.0, C	170.1, C			169.7, C
	20.5, CH$_3$	20.8, CH$_3$			20.8, CH$_3$
3'-OMe	62.4, CH$_3$	62.6, CH$_3$	56.5, CH$_3$	56.3, [d] CH$_3$	56.0, CH$_3$
5'-OMe				56.4, [d] CH$_3$	

[a] Measured at 150 MHz; [b] Measured at 175 MHz; [c] Data extracted from the HSQC (Heteronuclear Single Quantum Coherence) and HMBC spectra; [d] interchangeable.

Table 2. ^{1}H NMR data of compounds **1**, **2**, **7**, **8**, and **9** in CDCl$_3$.

No.	1 [a]	2 [a]	7 [b]	8 [a]	9 [b]
1	4.34, dd (11.0, 5.0); 3.96, dd (13.0, 11.0)	4.29, dd (11.0, 5.1); 3.93, dd (13.2, 11.0)	4.77, dd (10.5, 5.1); 3.84, dd (13.0, 10.5)	4.77, dd (10.6, 5.0); 3.84, dd (13.0, 10.6)	4.32, dd (10.8, 5.0); 3.93, dd (13.0, 10.8)
4	5.59, s	5.53, s	5.56, s	5.57, s	5.57, s
5	5.86, d (1.7)	5.79, d (1.9)	5.80, d (1.6)	5.80, d (1.8)	5.85, d (1.9)
8	6.11, d (10.7)	6.09, d (10.7)	4.74, d (10.7)	4.75, d (10.7)	6.11, d (10.7)
8a	2.98, dddd (13.0, 10.7, 5.0, 1.7)	2.94, dddd (13.2, 10.7, 5.1, 1.9)	2.88, dddd (13.0, 10.7, 5.1, 1.6)	2.88, dddd (13.0, 10.7, 5.0, 1.8)	2.96, dddd (13.0, 10.7, 5.0, 1.9)
9	1.56, s	1.56, s	1.47, s	1.47, s	1.54, s
10	5.91, dq (15.4, 1.3)	2.20, m	5.90, dq (15.4, 1.4)	5.90, dd (15.4, 1.5)	5.90, dd (15.3, 1.6)
11	6.45, dq (15.4, 7.0)	1.58, m	6.47, dq (15.4, 7.0)	6.46, dd (15.4, 7.0)	6.42, dd (15.3, 7.0)
12	1.88, dd (7.0, 1.3)	0.94, t (7.4)	1.87, dd (7.0, 1.4)	1.88, dd (7.0, 1.5)	1.87, dd (7.0, 1.6)
4'			6.52, s	6.38, s	6.23, s
6'					6.22, s
8'	2.50, s	2.50, s	2.48, s	2.51, s	2.39, s
8-OAc	2.17, s	2.16, s			2.15, s
3'-OMe	3.83, s	3.83, s	3.83, s	3.92, [c] s	3.74, s
5'-OH	6.04, s	6.04, s			
5'-OMe				3.89, [c] s	

[a] Measured at 600 MHz; [b] Measured at 700 MHz; [c] interchangeable.

The molecular formula of compound **2** was deduced as C$_{24}$H$_{26}$Cl$_2$O$_8$, containing two additional protons when compared to **1**. Investigation of the ^{1}H and ^{13}C NMR data of **2** (Tables 1 and 2) demonstrated the close structural similarity between **1** and **2** except for signals of the side chain at C-3. In particular, the resonances for the olefinic methine protons at δ_H 5.91 (H-10) and 6.45 (H-11) of **1** were replaced by two methylene groups at δ_H 2.20 (H$_2$-10) and 1.58 (H$_2$-11) in **2**. The COSY correlations between H$_2$-10/H$_2$-11/Me-12 together with the HMBC correlations from H$_2$-11 to C-3 (δ_C 168.2) and from H$_2$-10 to C-3 and C-4 (δ_C 100.9) confirmed the attachment of a *n*-propyl moiety at C-3 in **2**. The remaining substructure of **2** was determined to be identical to that of **1** by comparison of the 2D NMR data of **1** and **2**. The similar coupling constants and NOE relationships between **1** and **2** indicated that both share the same relative configuration.

When 3.5% NaCl as a constituent of the rice medium was replaced with 3.5% NaBr, a profound change in the metabolic profile of the fungus was observed. In total, two new brominated azaphilone derivatives (**7** and **8**), one new non-halogenated azaphilone (**9**), in addition to the known falconensins K (**10**) [23] and I (**11**) [23] were obtained.

The ^{1}H and ^{13}C NMR spectra of **7** (Tables 1 and 2) were comparable to those of the co-isolated known compound, falconensin K (**10**) [23]. The HRESIMS data of **7** established the molecular formula $C_{22}H_{23}BrO_7$, differing from that of **10** by the replacement of the chlorine with a bromine atom. The HMBC correlations from Me-8′ (δ_H 2.48) to C-2′ (δ_C 117.4), C-6′ (δ_C 105.3), and C-7′ (δ_C 137.4), from H-4′ to C-2′, C-6′, C-3′ (δ_C 156.2), and C-4′ (δ_C 154.0), from 3′-OMe (δ_H 3.83) to C-3′ together with the NOE correlation between 3′-OMe and Me-9 (δ_H 1.47) indicated the position of the bromine atom at C-6′ in **7**. The remaining azaphilone core structure including the relative configuration of **7** was confirmed to be identical to that of **10** after further inspection of the 2D NMR spectra of **7**. Thus, the structure of **7** was elucidated as shown (Figure 1), and the trivial name falconensin Q is proposed for this compound.

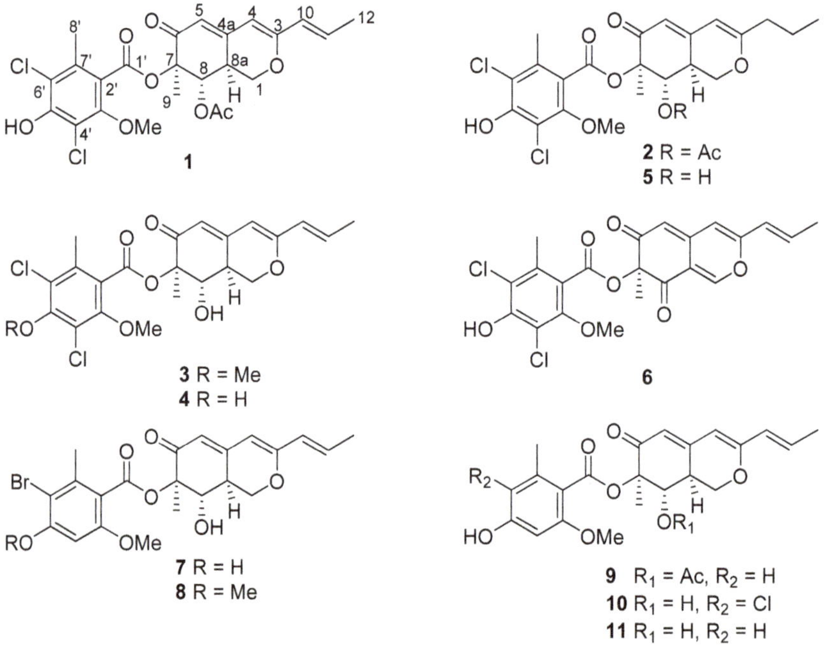

Figure 1. Structures of azaphilones isolated from *A. falconensis*.

Falconensin R (**8**) was isolated as a yellow oil. Its molecular formula was determined as $C_{23}H_{25}BrO_7$ by HRESIMS, containing an additional methyl group when compared to **7**. The NMR data of **8** (Tables 1 and 2) were similar to those of **7** except for the appearance of two methoxy groups at δ_C 56.4 and 56.3, and at δ_H 3.89 (5′-OMe) and 3.92 (3′-OMe) instead of one methoxy group in **7**. The location of the additional methoxy group at C-5′ in **8** was confirmed by the HMBC correlations from 3′-OMe to C-3′ (δ_C 157.3), from 5′-OMe to C-5′ (δ_C 155.8), and from H-4′ to C-2′ (δ_C 117.2), C-6′ (δ_C 106.6), C-3′ and C-5′, as well as based on the NOE correlations from H-4′ to 3′-OMe and 5′-OMe.

The molecular formula of **9** was determined as $C_{24}H_{26}O_8$ by HRESIMS, indicating 42 amu more than that of the co-isolated known falconensin I (**11**) ($C_{22}H_{24}O_7$) [23]. Comparison of the ^{1}H and ^{13}C

NMR spectra revealed that the structure of **9** was closely related to that of **11**, except for the appearance of an additional acetoxy group at δ_H 2.15 and δ_C 20.8 and 169.7 in **9**. The location of this additional acetoxy group at C-8 was confirmed by the COSY correlations between H-1ab/H-8a/H-8 and the HMBC correlations from H-8 to the carbon of the additional acetoxy group. The similar NOE correlations of **9** and **11** suggested that they shared the same configuration. Thus, compound **9** was elucidated as 8-*O*-acetyl analogue of falconensin I (**11**).

The specific optical rotation values of **1**, **2**, **7**, **8**, and **9** are positive (+99 − +233), which is in agreement with the positive values reported for other known falconensin derivatives with 7*R*, 8*S*, and 8a*S* configuration, suggesting that the new compounds share the same absolute configuration as the known derivatives [18,20,25]. Although the absolute configuration of other falconensin derivatives had been determined previously by Mosher's reaction and observation of Cotton effects of CD curves [18,20,25], no crystal structure had so far been reported for these compounds. To independently assign the absolute configuration of the isolated compounds, a single crystal X-ray diffraction analysis through anomalous dispersion of the known falconensin N (**5**), for which crystals of sufficient quality were obtained, was conducted. Herein, we present for the first time the crystal structure of falconensin N (**5**) (Figure 2). From the single-crystal structure refinement, the absolute structure assignment was based on the Flack parameter of 0.016(5) (Table S1) [26–29]. The crystal structure of falconensin N (**5**) is in agreement with the previously reported absolute configuration [18,20,25]. Based on these data, all falconensin derivatives isolated in this study are suggested to share the same 7*R*, 8*S*, and 8a*S* absolute configuration.

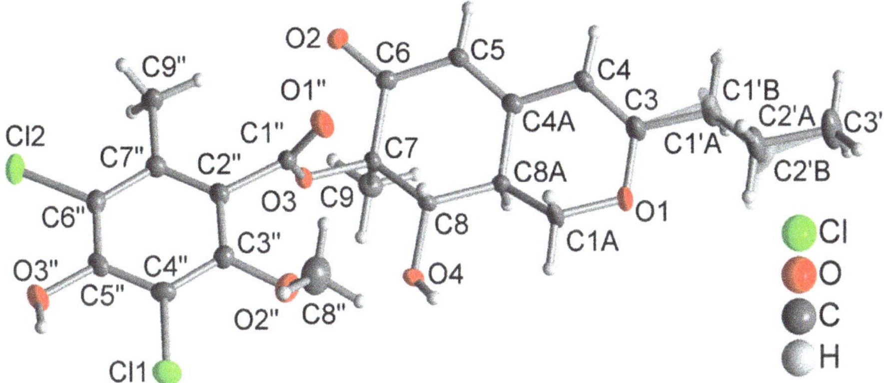

Figure 2. Molecular structure of **5** from a single-crystal X-ray structure determination (50% thermal ellipsoids; H atoms with arbitrary radii). The propyl group is disordered, and the two respective atom positions were refined with equal occupation factors.

Azaphilones constitute a class of fungal metabolites possessing various biological activities, including antiviral, antibacterial, antioxidant, hypolipidemic, cytotoxic, and anti-inflammatory properties [24,30]. For falconensins in particular, anti-inflammatory activity against 12-*O*-tetradecanoylphorbol-13-acetate-induced inflammatory ear edema in mice had been reported [21]. Compounds **1**, **3–9**, and **11** were evaluated for their anti-inflammatory activity in the triple negative breast cancer cell line NF-κB-MDA-MB-231. The IC_{50} values were calculated based on inhibition of the NF-κB-dependent luciferase activity and revealed NF-κB blockade for all compounds (Table 3). To exclude that cytotoxicity caused reduction of NF-κB activity, cell viability was determined in parallel. Compounds **1**, **4–7**, **9**, and **11** did not influence cell viability within the selected concentration range, whereas compound **8** reduced cell viability with an IC_{50} of 126.8 ± 5.4 µM, thus showing about 9-times higher potency in NF-κB-blockade compared to its cytotoxicity. Compound **3** with an IC_{50} of 89.7 ± 9.1 µM in the cytotoxicity assay only showed around 2-times higher potency in NF-κB-blockade.

Plotting the pIC_{50}, which is calculated as the negative decadic logarithm of IC_{50}, of the cell viability against the pIC_{50} of NF-κB inhibition, illustrates a greater anti-inflammatory potential of the examined compounds compared to their cytotoxic potential (Figure S40). In conclusion, inhibition of NF-κB signaling in the TNBC cell line MDA-MB-231 could be induced by **1**, **3 – 9**, and **11**, with **7** being the most potent compound.

Table 3. IC_{50} values (μM) of the compounds tested in MDA-MB-231 cells for NF-κB inhibition and inhibition of cell viability. [a]

Compound	NF-κB inhibition	Cell viability inhibition
Falconensin O (1)	15.7 ± 0.7	> 200
Falconensin A (3)	53.2 ± 21.4	89.7 ± 9.1
Falconensin M (4)	56.5 ± 8.3	> 200
Falconensin N (5)	71.0 ± 7.3	> 200
Falconensin H (6)	72.0 ± 28.1	> 400
Falconensin Q (7)	11.9 ± 2.1	> 200
Falconensin R (8)	14.6 ± 1.7	126.8 ± 5.4
Falconensin S (9)	20.1 ± 5.6	> 200
Falconensin I (11)	19.5 ± 2.5	> 400

[a] Average IC_{50} of at least three individual experiments ± SD tested in the concentration range of 400 μM to 0.78 μM (**3**, **6**, and **11**) and 200 μM to 0.78 μM (**1**, **4**, **5**, **7**, **8**, and **9**) in a two-fold serial dilution.

3. Materials and Methods

3.1. General Experimental Procedures

Optical rotations were measured using a Jasco P-2000 polarimeter. The compounds were dissolved in optically pure solvents Uvasol® (spectroscopic grade solvents, Merck). 1D and 2D NMR spectra were recorded on Bruker Avance III 300 or 600 or 700 MHz NMR spectrometers (Bruker BioSpin, Rheinstetten, Germany). Low-resolution mass spectra were recorded with an Ion-Trap-API Finnigan LCQ Deca (Thermo Quest) mass spectrometer, while high-resolution mass data were measured on a FTHRMS-Orbitrap (Thermo-Finnigan) mass spectrometer. HPLC analysis was conducted using a Dionex UltiMate-3400 SD with an LPG-3400SD pump coupled to a photodiode array detector (DAD3000RS) and employing a Knauer Eurospher C_{18} analytical column (125 × 4 mm i.d., 5 μm). Purification of the compounds was performed using semipreparative HPLC on the VWR Hitachi Chromaster HPLC system, 5160 Pump; 5410 UV detector; Eurosphere-100 C_{18}, 300 mm × 8 mm i.d., 10 μm, Knauer, Germany) with MeOH and H2O as eluents utilizing a flow rate of 5 mL/min. Column chromatography was performed using different stationary phases including Merck MN silica gel 60 M (0.04–0.063 mm), silica gel 60 RP-18 (40–63 μm), and Sephadex LH-20 (Merck). TLC plates precoated with silica gel 60 F_{254} (Merck) were used for monitoring fractions resulting from column chromatography. Detection of spots on the TLC was done by UV absorption at 254 and 365 nm followed by anisaldehyde spray reagent.

3.2. Fungal Material

The fungus was isolated from marine sediment which was collected at a depth of 25 m from the Canyon at Dahab, Red Sea, Egypt in November 2016. The fungus was identified as *A. falconensis* (GenBank accession No. MN905375) through amplification and sequencing of the internal transcribed spacer region including the 5.8S ribosomal DNA following by a subsequent BLAST search in NCBI as described before [31]. A deep-frozen specimen of the fungal strain has been deposited in one of the author's laboratory (P.P.).

3.3. Fermentation, Extraction, and Isolation

Initial fermentation of the fungus was conducted in ten 1L Erlenmeyer flasks. To each flask, 100 g of rice (Oryza Milchreis), 100 mL of demineralized water, and 3.5 g of NaCl were added. Thereafter, the flasks were autoclaved at 121 °C for 20 min and after cooling to room temperature, the fungus was inoculated on the rice medium. Fermentation of the fungus was continued under static conditions for 21 days at 20 °C until the fungus had totally overgrown the medium. Then, each flask was extracted with 600 mL EtOAc. After overnight soaking in EtOAc, the rice medium was cut into small pieces and shaken for 8 h at 150 rpm followed by evaporation of EtOAc, yielding around 16 g of EtOAc extract. For the OSMAC experiment, the same cultivation and extraction procedures of the initial cultivation were conducted except for the replacement of the added 3.5% NaCl with 3.5% NaBr and the cultivation on three flasks instead of ten. Eventually, the OSMAC experiment yielded approximately 2 g of EtOAc extract.

The initial crude extract of the large scale cultivation of the fungus obtained in presence of NaCl (16 g) was fractionated by vacuum liquid chromatography (VLC) on silica gel as a stationary phase utilizing a step gradient of solvents consisting of mixtures of *n*-hexane/EtOAc and CH_2Cl_2/MeOH, to yield 12 fractions (V1 to V12). Fraction V4 (874 mg) was further separated by Sephadex LH20 column chromatography using CH_2Cl_2-MeOH (1:1) as mobile phase affording eight subfractions (V4-S1 to V4-S8). Subfraction V4-S4 (162.8 mg) was purified by semi-preparative HPLC using acetonitrile-H_2O containing 0.1% formic acid (from 60:40 to 95:5 in 22 min) to afford **1** (1.8 mg), **2** (1.5 mg), and **3** (11.8 mg). Subfraction V4-S5 (139.6 mg) was further purified by semi-preparative HPLC using gradient elution with acetonitrile-H_2O containing 0.1% formic acid (60:40 to 95:5 in 22 min) affording **4** (6.3 mg) and **5** (2.2 mg). Fraction V6 (686.1 mg) was submitted to RP-VLC column using H_2O-MeOH gradient elution to yield 10 subfractions (V6-R1 to V6-R10). Subfraction V6-R7 (110 mg) was further purified using semi-preparative HPLC with a gradient of MeOH-H_2O containing 0.1% formic acid (67:33 to 80:20 in 20 min) to afford **6** (15.1 mg).

The EtOAc extract (2 g) obtained from the OSMAC experiment with addition of 3.5% NaBr, was also fractionated by VLC on silica gel as described before to give 13 fractions (BrV1 to BrV13). Subsequent purification of BrV5 (157 mg), BrV6 (32 mg), and BrV10 (140 mg) using Sephadex LH20 column chromatography with CH_2Cl_2-MeOH (1:1) as mobile phase and semi-preparative HPLC with a gradient of MeOH-H_2O containing 0.1% formic acid to afford **7** (3.1 mg), **8** (2.1 mg), **9** (1.2 mg), **10** (0.9 mg), and **11** (2.0 mg).

Falconensin O (**1**): Yellow oil; $[\alpha]_D^{20}$ +176 (*c* 0.2, MeOH); UV (MeOH) λ_{max} 355 and 206 nm; ^{1}H and ^{13}C NMR data, Tables 1 and 2; HRESIMS *m/z* 511.0923 [M + H]$^+$ (calcd for $C_{24}H_{25}Cl_2O_8$, 511.0921).

Falconensin P (**2**): Yellow amorphous powder; $[\alpha]_D^{20}$ +233 (*c* 0.2, MeOH); UV (MeOH) λ_{max} 329 and 216 nm; ^{1}H and ^{13}C NMR data, Tables 1 and 2; HRESIMS *m/z* 513.1073 [M + H]$^+$ (calcd for $C_{24}H_{27}Cl_2O_8$, 513.1077).

Falconensin Q (**7**): Yellow oil; $[\alpha]_D^{20}$ +182 (*c* 0.2, MeOH); UV (MeOH) λ_{max} 354 and 205 nm; ^{1}H and ^{13}C NMR data, Tables 1 and 2; HRESIMS *m/z* 479.0702 [M + H]$^+$ (calcd for $C_{22}H_{24}BrO_7$, 479.0700).

Falconensin R (**8**): Yellow oil; $[\alpha]_D^{20}$ +99 (*c* 0.2, MeOH); UV (MeOH) λ_{max} 353 and 206 nm; ^{1}H and ^{13}C NMR data, Tables 1 and 2; HRESIMS *m/z* 493.0853 [M + H]$^+$ (calcd for $C_{23}H_{26}BrO_7$, 493.0856).

Falconensin S (**9**): Yellow oil; $[\alpha]_D^{20}$ +105 (*c* 0.2, MeOH); UV (MeOH) λ_{max} 355 and 201 nm; ^{1}H and ^{13}C NMR data, Tables 1 and 2; HRESIMS *m/z* 443.1700 [M + H]$^+$ (calcd for $C_{24}H_{27}O_8$, 443.1700).

3.4. Crystallographic Analysis of Compound **5**

Crystals were obtained by solvent evaporation (MeOH). Data Collection: compound **5** was measured with a Bruker Kappa APEX2 CCD diffractometer with a microfocus tube and Cu-Kα radiation (λ = 1.54178 Å). APEX2 was used for data collection, SAINT for cell refinement and data reduction [32], and SADABS for experimental absorption correction [33]. The structure was solved by intrinsic phasing using SHELXT [34], and refinement was done by full-matrix least-squares on F^2 using SHELXL-2016/6 [35]. The hydrogen atoms were positioned geometrically (with C-H = 0.95 Å for aromatic CH, 1.00 Å for aliphatic CH, 0.99 Å for CH_2, and 0.98 Å for CH) and refined using riding models (AFIX 43, 13, 23, 137, respectively), with Uiso(H) = 1.2 Ueq(CH, CH_2), and 1.5 Ueq(CH_3). The absolute structure configuration of **5** was solved using anomalous dispersion from Cu-Kα, resulting in a Flack parameter of 0.016(5) using Parsons quotient method. All graphics were drawn using DIAMOND [36]. The structural data have been deposited in the Cambridge Crystallographic Data Center (CCDC No. 1976223).

3.5. Triple Negative Breast Cancer Studies

3.5.1. Cell Culture and Chemicals

Culture medium and supplements were purchased from Gibco (Fisher Scientific, Schwerte, Germany). Cell plates were obtained from Greiner bio-one (Frickenhausen, Germany). Cells were grown and incubated in a humidified 5% CO_2 atmosphere at constant 37 °C. The metastatic breast cancer cell line, MDA-MB-231, was purchased from the European Collection of Authenticated Cell Cultures (ECACC) (Salisbury, UK). Sub-culturing was performed in RPMI 1640 medium (#21875-034) supplemented with 15% (v/v) fetal calf serum (FCS) and 1% (v/v) penicillin-streptomycin (pen-strep) (10,000 U/mL). The MDA-MB-231 cell line stably expressing a plasmid with the NF-κB response element and the gene sequence for the firefly-luciferase protein (NF-κB- MDA-MB-231) was purchased from Signosis (Santa Clara, CA, USA; #SL-0043). For selection, 100 μg/mL hygromycin B (Life Technologies, Darmstadt, Germany; #10687010) was applied in high glucose DMEM (#41966-029) supplemented with 10% (v/v) FCS, 1% (v/v) pen-strep. Starvation medium for the NF-κB inhibition assay contained 1% (v/v) FCS, 1% (v/v) pen-strep, and 100 μg/mL hygromycin B. Cell detachment occurred by trypsinization in 0.25% trypsin-EDTA and cell counting was performed at 1:1 (v/v) dilution in Erythrosin B (BioCat, Heidelberg, Germany; #L13002) using the LUNA II automated cell counter (BioCat). Compounds **1**, **4**, **5**, **7**, **8**, and **9** were dissolved in dimethyl sulfoxide (DMSO) to a final concentration of 10 mM, whereas the compounds **3**, **6**, **11** were dissolved at 20 mM in DMSO. Further dilutions in cell culture medium contained maximal 2% DMSO.

3.5.2. NF-κB Inhibitory Assay

3×10^4 NFκB-MDA-MB-231 cells were seeded in total 100 μL medium per well in a 96-well plate (Greiner; #655098). On the next day, medium was exchanged, and triplicates were pre-incubated for 20 min without (negative control) or with compounds in total 100 μL starvation medium. Final concentration of the compounds ranged a twofold serial dilution starting from 400 μM (compound **3**, **6**, and **11**) or 200 μM (compound **1**, **4**, **5**, **7**, **8**, and **9**) to 0.78 μM. To activate NF-κB signaling, untreated cells (positive control) or compound treated cells were subsequently stimulated for 24 h with 20 ng/mL TNFα (Peprotech, Hamburg, Germany; #300-01A). Finally, cell lysis and measurement were done according to the manufacturer's instruction of the Luciferase Assay System (Promega, Mannheim, Germany; #E1500). Injection of equal volume of luciferase substrate with 10 s integration time and subsequent luminescence measurement was performed using the Spark® microplate reader (TECAN, Männedorf, Switzerland).

3.5.3. Cell Viability Assay

Using the CyBio® Well vario pipetting robot (Analytik Jena, Jena, Germany; #OL3381-24-730), 18 μL of the MDA-MB-231 cell suspension (2.8×10^5 cells/mL) were seeded per well in a 384-well plate (Greiner; #781074) and incubated for 24 h. The final concentration of the compounds ranged in twofold serial dilution steps starting from 400 μM (compound **3**, **6**, and **11**) or 200 μM (compound **1**, **4**, **5**, **7**, **8**, and **9**) to 0.78 μM. After 24 h compound stimulation in quadruples, cell lysis and measurement were done as prescribed in the manufacturer's instruction of the CellTiter-Glo® Luminescent Cell Viability Assay (Promega; #G7570). In short, it was applied equal volume of the CellTiter-Glo® reagent and luminescence was measured using the Spark® microplate reader (TECAN).

3.5.4. Statistical Analysis

Data of the NF-κB inhibitory assay and cell viability assay represent at least three individual experiments and were analyzed using GraphPad Prism (GraphPad Software, San Diego, USA; Version 8.1.2). For the NF-κB inhibitory assay, data below the relative light unit (RLU) of the negative control were excluded for further analysis. Half maximal inhibitory concentration (IC_{50}) values were determined by nonlinear regression analysis based on the dose–response inhibition calculation "log(inhibitor) vs. response–variable slope (four parameters)" without curve fitting.

Supplementary Materials: The following are available online at http://www.mdpi.com/1660-3397/18/4/204/s1, UV, HRMS, 1D and 2D NMR spectra of all the new compounds **1**, **2** and **7−9** as well as NF-κB inhibitory potential of the tested compounds and crystal data for compound **5**.

Author Contributions: Investigation, D.H.E.-K., F.S.Y., R.H., T.-O.K., C.J., W.L., I.R., and N.T.; writing—original draft preparation, D.H.E.-K.; writing—review and editing and supervision, Z.L. and P.P. All authors have read and agreed to the published version of the manuscript.

Funding: This project was supported by grants of the DFG (GRK 2158, project number 270650915) and the Manchot Foundation to P.P.

Acknowledgments: D.H.E. gratefully acknowledges the Egyptian Government (Ministry of Higher Education) for awarding a doctoral scholarship. Furthermore, we wish to thank Dr. Dent. Abdel Rahman O. El Mekkawi, EFR, PADI IDC staff instructor, founder of I Dive Tribe, Dahab, South Sinai – Egypt, for collecting the sediment sample.

Conflicts of Interest: The authors declare no conflict of interest.

References

1. Jimenez, C. Marine natural products in medicinal chemistry. *ACS Med. Chem. Lett.* **2018**, *9*, 959–961. [CrossRef] [PubMed]
2. Blunt, J.W.; Carroll, A.R.; Copp, B.R.; Davis, R.A.; Keyzers, R.A.; Prinsep, M.R. Marine natural products. *Nat. Prod. Rep.* **2018**, *35*, 8–53. [CrossRef]
3. Wang, Y.T.; Xue, Y.R.; Liu, C.H. A brief review of bioactive metabolites derived from deep-sea fungi. *Mar. Drugs* **2015**, *13*, 4594–4616. [CrossRef] [PubMed]
4. Fenical, W.; Jensen, P.R.; Cheng, X.C. Halimide, A Cytotoxic Marine Natural Product, and Derivatives Thereof. U.S. Patent No. 6,069,146, 30 May 2000.
5. Petersen, L.E.; Kellermann, M.Y.; Schupp, P.J. Secondary metabolites of marine microbes: From natural products chemistry to chemical ecology. In *YOUMARES 9-The oceans: Our Research, our Future*; Jungblut, S., Liebich, V., Bode-Dalby, M., Eds.; Springer International Publishing: Cham, Switzerland, 2020; pp. 159–180.
6. Available online: https://www.beyondspringpharma.com/ChannelPage/index.aspx (accessed on 6 November 2019).
7. Zhang, Z.; He, X.; Wu, G.; Liu, C.; Lu, C.; Gu, Q.; Che, Q.; Zhu, T.; Zhang, G.; Li, D. Aniline-tetramic acids from the deep-sea-derived fungus *Cladosporium sphaerospermum* L3P3 cultured with the HDAC inhibitor SAHA. *J. Nat. Prod.* **2018**, *81*, 1651–1657. [CrossRef] [PubMed]
8. Hertweck, C. Hidden biosynthetic treasures brought to light. *Nat. Chem. Biol.* **2009**, *5*, 450–452. [CrossRef]
9. Daletos, G.; Ebrahim, W.; Ancheeva, E.; El-Neketi, M.; Lin, W.; Chaidir, C. Microbial co-culture and OSMAC approach as strategies to induce cryptic fungal biogenetic gene clusters. In *Chemical Biology of Natural Products*; Grothaus, P., Cragg, G.M., Newman, D.J., Eds.; CRC press: Boca Raton, FL, USA, 2017; pp. 233–284.
10. Hofs, R.; Walker, M.; Zeeck, A. Hexacyclinic acid, a polyketide from *Streptomyces* with a novel carbon skeleton. *Angew. Chem. Int. Ed.* **2000**, *39*, 3258–3261. [CrossRef]
11. Zhang, M.; Wang, W.L.; Fang, Y.C.; Zhu, T.J.; Gu, Q.Q.; Zhu, W.M. Cytotoxic alkaloids and antibiotic nordammarane triterpenoids from the marine-derived fungus *Aspergillus sydowi*. *J. Nat. Prod.* **2008**, *71*, 985–989. [CrossRef]
12. Saleem, M.; Ali, M.S.; Hussain, S.; Jabbar, A.; Ashraf, M.; Lee, Y.S. Marine natural products of fungal origin. *Nat. Prod. Rep.* **2007**, *24*, 1142–1152. [CrossRef]
13. Ge, H.M.; Yu, Z.G.; Zhang, J.; Wu, J.H.; Tan, R.X. Bioactive alkaloids from endophytic *Aspergillus fumigatus*. *J. Nat. Prod.* **2009**, *72*, 753–755. [CrossRef]

14. Frank, M.; Ozkaya, F.C.; Muller, W.E.G.; Hamacher, A.; Kassack, M.U.; Lin, W.; Liu, Z.; Proksch, P. Cryptic secondary metabolites from the sponge-associated fungus *Aspergillus ochraceus*. *Mar. Drugs* **2019**, *17*, 99. [CrossRef]

15. El-Kashef, D.H.; Daletos, G.; Plenker, M.; Hartmann, R.; Mandi, A.; Kurtan, T.; Weber, H.; Lin, W.; Ancheeva, E.; Proksch, P. Polyketides and a dihydroquinolone alkaloid from a marine-derived strain of the fungus *Metarhizium marquandii*. *J. Nat. Prod.* **2019**, *82*, 2460–2469. [CrossRef] [PubMed]

16. Kuppers, L.; Ebrahim, W.; El-Neketi, M.; Ozkaya, F.C.; Mandi, A.; Kurtan, T.; Orfali, R.S.; Muller, W.E.G.; Hartmann, R.; Lin, W.H.; et al. Lactones from the sponge-derived fungus *Talaromyces rugulosus*. *Mar. Drugs* **2017**, *15*, 359. [CrossRef] [PubMed]

17. Samson, R.A.; Visagie, C.M.; Houbraken, J.; Hong, S.B.; Hubka, V.; Klaassen, C.H.; Perrone, G.; Seifert, K.A.; Susca, A.; Tanney, J.B.; et al. Phylogeny, identification and nomenclature of the genus *Aspergillus*. *Stud. Mycol.* **2014**, *78*, 141–173. [CrossRef]

18. Itabashi, T.; Nozawa, K.; Miyaji, M.; Udagawa, S.; Nakajima, S.; Kawai, K. Falconensins A, B, C, and D, new compounds related to azaphilone from *Emericella falconensis*. *Chem. Pharm. Bull.* **1992**, *40*, 3142–3144. [CrossRef]

19. Itabashi, T.; Nozawa, K.; Nakajima, S.; Kawai, K. A new azaphilone, falconensin H, from *Emericella falconensis*. *Chem. Pharm. Bull.* **1993**, *41*, 2040–2041. [CrossRef]

20. Itabashi, T.; Ogasawara, N.; Nozawa, K.; Kawai, K. Isolation and structures of new azaphilone derivatives, falconensins E−G, from *Emericella falconensis* and absolute configurations of falconensins A−G. *Chem. Pharm. Bull.* **1996**, *44*, 2213–2217. [CrossRef]

21. Yasukawa, K.; Itabashi, T.; Kawai, K.; Takido, M. Inhibitory effects of falconensins on 12-*O*-tetradecanoylphorbol-13-acetate-induced inflammatory ear edema in mice. *J. Nat. Med.* **2008**, *62*, 384–386. [CrossRef]

22. Agrawal, A.K.; Pielka, E.; Lipinski, A.; Jelen, M.; Kielan, W.; Agrawal, S. Clinical validation of nuclear factor kappa B expression in invasive breast cancer. *Tumour Biol.* **2018**, *40*. [CrossRef]

23. Ogasawara, N.; Kawai, K.I. Hydrogenated azaphilones from *Emericella falconensis* and *E. fruticulosa*. *Phytochemistry* **1998**, *47*, 1131–1135. [CrossRef]

24. Gao, J.M.; Yang, S.X.; Qin, J.C. Azaphilones: Chemistry and biology. *Chem. Rev.* **2013**, *113*, 4755–4811. [CrossRef]

25. Huang, H.; Wang, F.; Luo, M.; Chen, Y.; Song, Y.; Zhang, W.; Zhang, S.; Ju, J. Halogenated anthraquinones from the marine-derived fungus *Aspergillus* sp. SCSIO F063. *J. Nat. Prod.* **2012**, *75*, 1346–1352. [CrossRef]

26. Flack, H.D. On enantiomorph-polarity estimation. *Acta Crystallogr. Sect. A* **1983**, *39*, 876–881. [CrossRef]

27. Flack, H.D.; Bernardinelli, G. Absolute structure and absolute configuration. *Acta Crystallogr. Sect. A* **1999**, *55*, 908–915. [CrossRef] [PubMed]

28. Flack, H.D.; Bernardinelli, G. The use of X-ray crystallography to determine absolute configuration. *Chirality* **2008**, *20*, 681–690. [CrossRef] [PubMed]

29. Flack, H.D.; Sadki, M.; Thompson, A.L.; Watkin, D.J. Practical applications of averages and differences of Friedel opposites. *Acta Crystallogr. Sect. A* **2011**, *67*, 21–34. [CrossRef]

30. Osmanova, N.; Schultze, W.; Ayoub, N. Azaphilones: A class of fungal metabolites with diverse biological activities. *Phytochem. Rev.* **2010**, *9*, 315–342. [CrossRef]

31. Kjer, J.; Debbab, A.; Aly, A.H.; Proksch, P. Methods for isolation of marine-derived endophytic fungi and their bioactive secondary products. *Nat. Protoc.* **2010**, *5*, 479–490. [CrossRef]

32. Bruker. *Bruker AXS: Apex2, data collection program for the CCD area-detector system; SAINT, data reduction and frame integration program for the CCD area-detector system*; Bruker: Billerica, MA, USA, 2014–2015.

33. Sheldrick, G.M. *SADABS: Area-Detector Absorption Correction*; University of Goettingen: Goettingen, Germany, 1996.

34. Sheldrick, G.M. SHELXT–Integrated space-group and crystal-structure determination. *Acta Crystallogr. Sect. A Found. Adv.* **2015**, *71*, 3–8. [CrossRef]

35. Sheldrick, G.M. Crystal structure refinement with SHELXL. *Acta Crystallogr. Sect. C Struct. Chem.* **2015**, *71*, 3–8. [CrossRef]

36. Brandenburg, K. *Diamond (Version 4), Crystal and Molecular Structure Visualization*; Crystal Impact–K, Brandenburg & H. Putz Gbr: Bonn, Germany, 2009.

marine drugs

Article

Antimicrobial and Antibiofilm Activities of the Fungal Metabolites Isolated from the Marine Endophytes *Epicoccum nigrum* M13 and *Alternaria alternata* 13A

M. Mallique Qader [1,2,†], Ahmed A. Hamed [3,†], Sylvia Soldatou [4], Mohamed Abdelraof [3], Mohamed E. Elawady [5], Ahmed S. I. Hassane [1,6], Lassaad Belbahri [7], Rainer Ebel [4,*] and Mostafa E. Rateb [1,*]

1   School of Computing, Engineering, & Physical Sciences, University of the West of Scotland, Paisley PA1 2BE, UK; mallique@uic.edu (M.M.Q.); Ahmedsayed.hassane@nhs.scot (A.S.I.H.)
2   National Institute of Fundamental Studies, Hantana Road, Kandy 20000, Sri Lanka
3   National Research Centre, Microbial Chemistry Department, 33 El-Buhouth Street, Dokki, Giza 12622, Egypt; aa.shalaby@nrc.sci.eg (A.A.H.); mr.mahmoud@nrc.sci.eg (M.A.)
4   Marine Biodiscovery Centre, Department of Chemistry, University of Aberdeen, Aberdeen AB24 3UE, UK; s.soldatou@rgu.ac.uk
5   National Research Centre, Microbial Biotechnology Department, 33 El-Buhouth Street, Dokki, Giza 12622, Egypt; ms.el-awady@nrc.sci.eg
6   Aberdeen Royal Infirmary, Foresterhill Health Campus, Aberdeen AB25 2ZN, UK
7   Laboratory of Soil Biology, University of Neuchatel, 2000 Neuchatel, Switzerland; lassaad.belbahri@unine.ch
*   Correspondence: r.ebel@abdn.ac.uk (R.E.); Mostafa.Rateb@uws.ac.uk (M.E.R.); Tel.: +44-1224-272930 (R.E.); +44-141-8483072 (M.E.R.)
†   Both authors contributed equally to the work.

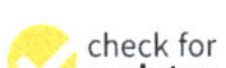
check for updates

Citation: Qader, M.M.; Hamed, A.A.; Soldatou, S.; Abdelraof, M.; Elawady, M.E.; Hassane, A.S.I.; Belbahri, L.; Ebel, R.; Rateb, M.E. Antimicrobial and Antibiofilm Activities of the Fungal Metabolites Isolated from the Marine Endophytes *Epicoccum nigrum* M13 and *Alternaria alternata* 13A. *Mar. Drugs* **2021**, *19*, 232. https://doi.org/10.3390/md19040232

Academic Editor: Vassilios Roussis

Received: 24 March 2021
Accepted: 19 April 2021
Published: 20 April 2021

Publisher's Note: MDPI stays neutral with regard to jurisdictional claims in published maps and institutional affiliations.

Abstract: Epicotripeptin (**1**), a new cyclic tripeptide along with four known cyclic dipeptides (**2–5**) and one acetamide derivative (**6**) were isolated from seagrass-associated endophytic fungus *Epicoccum nigrum* M13 recovered from the Red Sea. Additionally, two new compounds, cyclodidepsipeptide phragamide A (**7**) and trioxobutanamide derivative phragamide B (**8**), together with eight known compounds (**9–16**), were isolated from plant-derived endophyte *Alternaria alternata* 13A collected from a saline lake of Wadi El Natrun depression in the Sahara Desert. The structures of the isolated compounds were determined based on the 1D and 2D NMR spectroscopic data, HRESIMS data, and a comparison with the reported literature. The absolute configurations of **1** and **7** were established by advanced Marfey's and Mosher's ester analyses. The antimicrobial screening indicated that seven of the tested compounds exhibited considerable (MIC range of 2.5–5 µg/mL) to moderate (10–20 µg/mL) antibacterial effect against the tested Gram-positive strains and moderate to weak (10–30 µg/mL) antibacterial effect against Gram-negative strains. Most of the compounds exhibited weak or no activity against the tested Gram-negative strains. On the other hand, four of the tested compounds showed considerable antibiofilm effects against biofilm forming Gram-positive and Gram-negative strains.

Keywords: epicotripeptin; phragamide; *Epicoccum*; *Alternaria*; antimicrobial; antibiofilm

1. Introduction

Marine microorganisms are considered a rich source for drugs, drug leads, and agrochemicals [1–4]. Fungi isolated from different marine environments have been widely studied since they are cosmopolitan organisms that can survive in all phases of marine habitats regardless of environmental conditions. As a result, complex secondary metabolites are produced by marine fungi and analysis of their chemical profiles provides structurally diverse metabolites with new and novel chemical scaffolds, as well as promising biological activities [5,6].

Alternaria and *Epicoccum* are two ubiquitous fungal genera that are widely distributed in Nature [7,8]. They are found as endophytic, pathogenic, and saprophytic fungi [8,9].

Alternaria spp. are known as opportunistic pathogens that cause more than 300 plant diseases, having a detrimental impact on the agricultural economy [10]. The literature indicated that different alkaloids, terpenoids, steroids, phenolics, quinones, and pheromones are biosynthesised by more than one *Alternaria* species [11]. These metabolites are known to have potential phytotoxic, cytotoxic, antifungal and antimicrobial, and anticancer activities [11]. Meanwhile, *Epicoccum* spp. have been isolated as potential endophytic fungi residing in the sugarcane plant, which significantly impacts high root biomass and controls sugarcane pathogens [12]. Similarly, *Epicoccum* spp. are well known for their biocontrol activity against host pathogens, especially in sunflower, cotton, apple, and peaches [12]. Chemically, *Epicoccum* spp. have been shown to produce a rich number of carotenoid pigments, heterocyclic compounds, sulphur compounds, phenolics, and polysaccharides, which exhibit biologically important activities such as antioxidant, antimicrobial, herbicidal, antiviral, phytotoxic, anticancer, antitumor, and immunosuppressive [9,13–17]. Moreover, *E. nigrum* produces a range of diketopiperazine or cyclic dipeptides, which showed potential bioactivities such as anti-HIV, antifungal, and antibacterial [14,15]. The *Epicoccum*-derived metabolites are characterised by the presence of sulphur bridges, which possess diverse pharmacological effects. Though sulphur-containing metabolites are frequently found in Nature, the secondary metabolites with cross–ring sulphur bridges and S-methyl groups are rare and mainly reported from fungi [13].

In 2017, we initiated a collaborative project between Egypt and the UK aiming at the isolation of new endophytic fungal strains from different under-explored marine habitats in Egypt to be screened for their antimicrobial effect, with the ultimate aim of incorporating their bioactive extracts or metabolites into textiles used in Egyptian hospitals to reduce nosocomial infections. As a result, 21 out of 32 endosymbiotic marine-derived fungal isolates were isolated from both Hurghada, Red Sea and Wadi El-Natrun depression in Sahara, identified by molecular biological traits, and selected for further study based on their antimicrobial effects [18]. In this study, the chemical profiles following large scale fermentation of two endophytic fungal isolates, *Epicoccum nigrum* M13 isolated from the seagrass *Thalassia hemprichii* collected from Hurghada, Red sea and *Alternaria alternata* 13A recovered from the leaves of *Phragmites australis* collected from Wadi El-Natrun depression were studied. Herein, we report the large-scale fermentation, isolation, structure elucidation of new and known fungal metabolites of different chemical classes and their antimicrobial and antibiofilm activities on a panel of various pathogenic organisms.

2. Results and Discussion

2.1. Isolation and Identification of the Endophytic Fungal Strains

Out of the 21 endosymbiotic marine-derived fungal isolates recovered and prioritized in our previous study, we have selected the isolates M13 and 13A for scale up due to their promising antimicrobial effects and prolific chemical profiles established by the LC-HRMS analysis. The fungal strains M13 and 13A were isolated from the marine seagrass *Thalassia hemprichii* (collected from Makady bay, south Hurghada, Egypt) and the plant *Phragmites australis* (collected from Lake El-Bida, Wadi El-Natrun, Egypt) and identified using morphological features and genetic markers (ITS rDNA and β-tubulin) as *Epicoccum nigrum* and *Alternaria alternata*, respectively, as described previously [18].

2.2. Fermentation, Isolation, and Structure Elucidation

Large scale fermentation of the pure culture of *E. nigrum* M13 was performed on a modified marine ISP2 medium. The total extract was prepared using methanol (MeOH), defatted with *n*-hexane, and sequentially extracted with dichloromethane (DCM) and ethyl acetate (EtOAc). Both DCM and EtOAc fractions were analysed for their metabolite profiles using analytical RP–HPLC equipped with a diode array UV detector. The antimicrobial screening indicated the DCM fraction as the active subfraction, and further purification was carried out using RP-HPLC analysis which yielded six compounds (**1–6**, Figure 1).

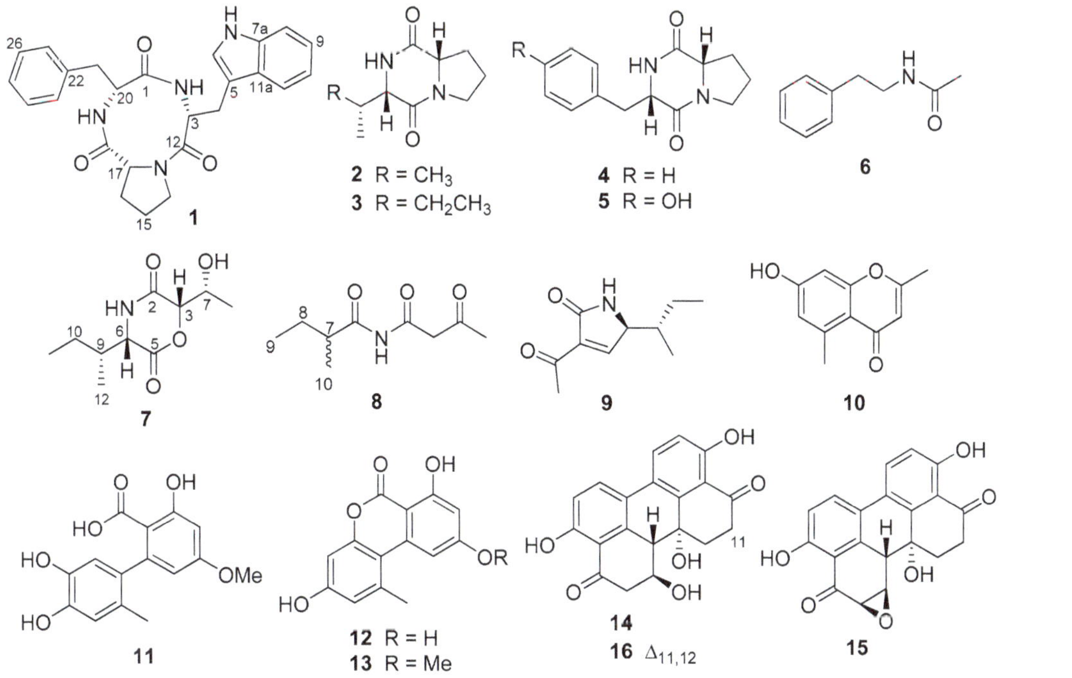

Figure 1. Structures of the fungal metabolites isolated from the *E. nigrum* M13 (**1–6**) and *A. alternata* 13A (**7–16**).

Epicotripeptin (**1**) was isolated from the DCM fraction as pale yellow coloured amorphous solid. The molecular formula $C_{25}H_{26}N_4O_3$, indicating fifteen degrees of unsaturation, was determined by the HRESIMS analysis which showed a quasimolecular ion peak at *m/z* 453.1899 [M + Na]$^+$. NMR data of compound **1** (Table 1) indicated the presence of 25 carbons which included five diastereotopic methylenes, three methines, ten aromatic/olefinic protons, four quaternary carbons, and three carbonyl carbons. The molecular formula together with the distinct ^{1}H and ^{13}C NMR profile indicated a peptide molecule. Analysis of 1D and 2D NMR data strongly suggested the presence of the amino acids phenylalanine (monosubstituted benzene ring at δ_H 7.26/H-23/27, 7.25/H-24/26, 7.19/H-25), tryptophan (1,2-disubstituted benzene ring, δ_H 7.36/H-8, 7.05/H-9, 6.96/H-10, 7.56/H-11, 7.17/H-6), and proline (δ_H 3.26–3.38/H$_2$-14, 1.61–1.71/H$_2$-15, 1.39–1.98/H$_2$-16, 10.85, 7-NH). This assignment was corroborated by the COSY spectrum, while HMBC correlations from 7-NH to C-5, H$_2$-4 to C-3 and C-5, H-3 to C-12, H-17 to C-18, H-23/25 to C-21, H$_2$-21 to C-1, as well as 19-NH to C-17 and C-1 allowed for connecting the carbonyl groups and the aliphatic moieties with the respective aromatic substructures and established the sequence of the amino acids within the tripeptide (Figure 2A).

The absolute configuration of the three amino acid residues of **1** was established using the advanced Marfey's method [19], which upon comparison with authentic standards revealed all of them to be present as their respective L-forms. Therefore, the absolute configuration of **1** was established as cyclo(L-Trp-L-Pro-L-Phe), a new natural product for which the name epicotripeptin is proposed.

Table 1. ^{1}H (600 MHz) and ^{13}C (150 MHz) NMR spectroscopic data for **1** (DMSO–d_6, 298 K).

Position	^{1}H (Mult., J in Hz)	^{13}C, Mult.	HMBC
Tryptophan moiety			
1-CO		165.1, C	
2-NH	7.73 (br s)		C-3, C-4, C-12, C-17, C-18
3	4.30 (t, 5.4)	55.2, CH	C-4, C-5, C-12, C-18
4	3.24–3.07 (m)	25.8, CH$_2$	C-3, C-5, C-6, C-12, C-11a
5		109.4, C	
6	7.17 (m)	124.3, CH	C-5, C-7a, C-11a
7-NH	10.85 (br s)		C-5, C-6, C-7a, C-11a
7a		136.0, C	
8	7.32 (d, 7.9)	111.2, CH	C-10, C-11a
9	7.05 (t, 7.6)	120.8, CH	C-11, C-7a
10	6.96 (t, 7.6)	118.2, CH	C-8, C-11a
11	7.56 (d, 7.8)	118.6, CH	C-5, C-9, C-7a, C-11a
11a		127.4, C	
Proline moiety			
12-CO		165.5, C	
14	3.38–3.26 (m)	44.5, CH$_2$	C-15, C-16, C-17
15	1.71–1.61 (m)	21.8, CH$_2$	C-14, C-16, C-17
16	1.98–1.39 (m)	27.7, CH$_2$	C-14, C-15, C-17, C-18
17	4.08 (m)	58.4, CH	C-16, C-18
Phenylalanine moiety			
18-CO		168.9, C	
19-NH	7.97 (br s)		C-1, C-17, C-18, C-20, C-21
20	4.34 (t, 5.4)	55.7, CH	C-1, C-18, C-21, C-22
21	3.03 (dd, 14.6, 4.9)	35.4, CH$_2$	C-1, C-20, C-22, C-23/27
22		137.3, C	
23/27	7.26 (d, 6.8)	129.8, CH	C-21, C-25
24/26	7.25 (t, 6.8)	127.9, CH	C-22
25	7.19 (m)	126.3, CH	C-23/C-27

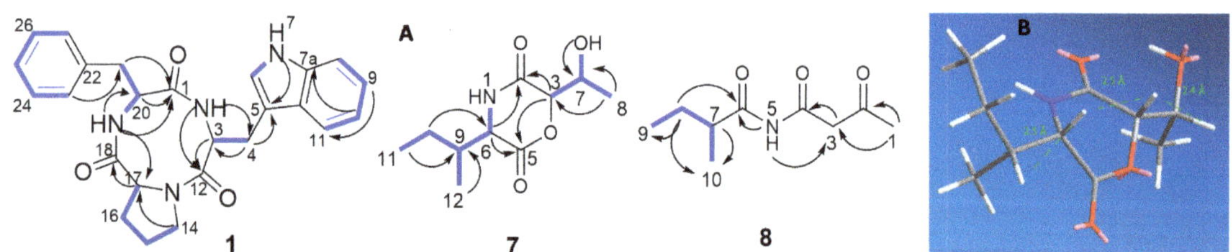

Figure 2. (A) Key COSY ▬ and HMBC ⌒ correlations of compounds **1**, **7**, and **8**; (B) molecular mechanics simulations of compound **7**.

The other isolated compounds from *E. nigrum* M13 were identified as the known cyclo(L-Pro-L-Val) (**2**) [20], cyclo(L-Pro-L-Ile) (**3**) [20], cyclo(L-Pro-L-Phe) (**4**) [21], cyclo(L-Pro-L-Tyr) (**5**) [21], and *N*- (2-phenylethyl)acetamide (**6**) [22] (Figure 1) based on their accurate mass analysis, NMR spectral, and other physical data in comparison with data reported in the literature.

Similarly, large scale fermentation of the pure culture of *A. alternata* 13A was performed on a rice medium prepared in seawater, followed by extraction and fractionation steps. Antimicrobial screening indicated that both DCM and EtOAc fractions were active, thus both were selected for further processing. Purification of the DCM and EtOAc fractions using different chromatographic techniques and finally using RP-HPLC yielded a total of 10 fungal metabolites (**7–16**).

Compound **7**, obtained as pale yellow amorphous solid, had the molecular formula of C$_{10}$H$_{17}$NO$_4$ with three degrees of unsaturation as determined by HRESIMS (*m/z* 214.1082

[M – H]$^-$). The NMR data showed that the molecule contained three methyl, four methine, one diastereotopic methylene, and two carbonyl carbons (Table 2). Inspection of the NMR data revealed the presence of an isoleucine moiety in **7**, similar to the known compound tenuazonic acid (**9**), which was confirmed by the COSY spin system including H$_3$-12, H$_3$-11, H$_2$-10, H-9, H-6, and the NH signal. Similarly, the COSY spectrum also indicated the presence of a 2-hydroxypropyl moiety, as evident from the spin system comprising δ_H 4.65, (H-3), δ_H 4.05 (H-7), δ_H 5.45 (OH), and δ_H 1.14 (H$_3$-8) (Figure 2A). Two carbonyl groups at δ_C 168.1 (C-2) and δ_C 166.0 (C-5) were connected to these two substructures based on their correlations in the HMBC spectrum (Figure 2A), clearly establishing the planar structure of **7** as 3-(sec-butyl)-6-(1-hydroxyethyl)morpholine-2,5-dione, a new natural product for which the trivial name phragamide A is proposed.

Table 2. ^{1}H (600 MHz) and ^{13}C (150 MHz) NMR spectroscopic data for **7** and **8** (DMSO–d_6, 298 K).

Position	Phragamide A (7)		Phragamide B (8)	
	δ_H (Mult., *J* in Hz)	δ_C, Mult.	δ_H (Mult., *J* in Hz)	δ_C, Mult.
1			2.15 (s)	30.0, CH$_3$
2		168.1, C		201.9, C
3	4.65 (s)	81.8, CH	3.76 (br s)	52.9, CH$_2$
4				169.1, C
5		166.0, C		
6	4.10 (br s)	57.5, CH		176.7, C
7	4.05 (m)	68.4, CH	2.50 (m)	41.8, CH
8	1.14 (d, 6.2)	18.6, CH$_3$	1.51–1.33 (m)	26.3, CH$_2$
9	1.93 (m)	38.3, CH	0.81 (t, 7.6)	10.9, CH$_3$
10	1.41–1.28 (m)	24.5, CH$_2$	0.98 (d, 6.9)	16.2, CH$_3$
11	0.87 (t, 7.2)	11.8, CH$_3$		
12	0.92 (d, 7.5)	14.7, CH$_3$		
1-NH	8.36 (br s)			
5-NH			10.76 (br s)	
7-OH	5.45 (brs)			

To determine the absolute configuration of the secondary alcohol at C-7, the modified Mosher ester reaction [23] was carried out. Positive $\Delta\delta$ were observed for the H-3, H-6, H-9, H-10, H-11, and H-12, whereas negative $\Delta\delta$ were observed in H-8 (Table S1), which allowed us to assign the absolute configuration as R. Based on the small coupling constant between H-3 and H-7, a syn-configuration was proposed with the chiral centre at C-3, which was thus judged to have S-configuration. The advanced Marfey's method [19] was carried out to establish the absolute configuration of the isoleucine residue. The retention time for the FDLA-derivatized amino acid residue of **7** was compared to that of the FDLA-derivatized standard amino acid, allowing the assignment of an L-configuration (6S,9R) to the isoleucine residue. This was further confirmed by the strong ROESY correlations between H-3/H-7, H-3/H-6, and H-6/H-9, as well as the molecular mechanics simulations (Figure 2A). It is worth noting that phragamide A (**7**) is a cyclodidepsipeptide belonging to the rare dioxomorpholine class of compounds and is closely related to three fungal metabolites previously characterised from Hypericum barbatum-derived endophyte Fusarium sporotrichioides [24].

Compound **8** was obtained from the DCM fraction as brownish oil. HRESIMS showed a quasimolecular ion peak at *m/z* 184.0979 [M – H]$^-$ indicating the molecular formula of C$_9$H$_{15}$NO$_3$ with three degrees of unsaturation. ^{1}H and ^{13}C NMR spectra showed that the molecule contained one methine, two methylene, and three methyl groups together with three carbonyl groups and an NH proton. From the chemical shift values and the multiplicities together with the COSY spin system observed, a 2-methylbutanoyl moiety was readily identified (Table 2), similar to compound **7** (vide supra). Analysis of the remaining signals, including HMBC correlations (Figure 2A) established the planar structure of **8** as 2-methyl-*N*-(3-oxobutanoyl) butanamide, a new compound for which the trivial name phragamide B is proposed. No attempt was made to establish the absolute configuration of **8** due to the scarcity of material obtained. It is interesting to note that **8** displays a degree of chemical similarity to the well-known tenuazonic acid (**9**), which was likewise detected in the same culture of *A. alternata* 13A.

The other isolated compounds from *A. alternata* 13A were identified as the known tenuazonic acid (**9**) [25], altechromone A (**10**) [26], altenusin (**11**) [27], alternariol (**12**) [28], alternariol monomethylether (**13**) [26], altertoxin I (**14**), altertoxin II (**15**) [29], and alter-perylenol (**16**) [30] (Figure 1) based on the comparison of their accurate masses, NMR spectra, and optical rotation data with literature values.

2.3. Biological Activity

2.3.1. Antimicrobial Activity

The rapid development of antimicrobial resistance is considered one of the major health concerns, with pathogenic microorganisms increasingly becoming resistant to antimicrobial or anti-infective agents. If left unattended, diseases associated with drug-resistant pathogens could kill more people than cancer [31]. For the discovery of new antimicrobial agents from natural sources that could be efficient in the incorporation of textiles, we have performed initial antimicrobial screening which indicated that the total extract of each of M13 and 13A exhibited promising antimicrobial effects against a panel of pathogenic strains tested [18]. The antimicrobial activity of the pure compounds (**1–6**) of *E. nigrum* M13 and compounds (**7–16**) of *A. alternata* 13A was assessed against a panel of pathogenic microorganisms comprising Gram-positive bacteria (*Staphylococcus aureus* and *Bacillus subtilis*), Gram-negative bacteria (*Escherichia coli, Pseudomonas aeruginosa, Klebsiella pneumonia, and Proteus vulgaris*), and yeast (*Candida albicans*) (Table 3). Epicotripeptin (**1**) showed considerable activity against both Gram-positive tested bacteria, while a moderate to weak inhibition was observed against all tested Gram-negative bacteria and *C. albicans*. Cyclo(L-Pro-L-Val) (**2**) and cyclo(L-Pro-L-Ile) (**3**) displayed a weak antibacterial activity only against Gram-positive bacteria, while cyclo(L-Pro-L-Phe) (**4**) and cyclo(L-Pro-L-Tyr) (**5**) displayed a moderate antibacterial activity against Gram-positive bacteria, weak activity against Gram-negative bacteria, and no antifungal activity.

Table 3. Minimum inhibitory concentrations (MIC) of the pure compounds (**1–16**) isolated from *E. nigrum* M13 and *A. alternata* 13A against bacterial and fungal pathogens.

Compound	Minimum Inhibitory Concentration (MIC, µg/mL) *						
	S. aureus	*B. subtilis*	*E. coli*	*K. pneumonia*	*P. vulgaris*	*P. aeruginosa*	*C. albicans*
1	2.5	2.5	10	20	10	30	30
2	50	40	-	-	-	-	-
3	20	20	-	-	-	-	-
4	10	10	30	30	30	-	-
5	10	10	30	30	30	-	-
6	40	40	-	-	-	-	40
7	5	5	20	-	-	10	20
8	30	40	30	40	40	-	50
9	10	10	-	-	-	-	-
10	-	-	-	-	-	-	-
11	20	20	-	-	-	30	50
12	20	30	-	-	-	40	50
13	30	30	-	-	-	-	40
14	30	30	-	-	-	-	40
15	30	20	-	-	-	-	40
16	30	20	-	-	-	-	30
Cip	0.62	0.31	1.25	1.25	1.25	2.5	-
Nys	-	-	-	-	-	-	5

* The average of two independent replicates, positive controls: Cip: Ciprofloxacin; Nys: Nystatin; -: not detected.

Previous reports indicated that the diketopiperazine (DKP) derivatives isolated from different microbial sources exhibited antibacterial [32] and antifungal [33] activities. Li et al. isolated two cyclic dipeptides, cyclo(D-Pro-L-Tyr), and cyclo(L-Pro-L-Tyr) from *Lactobacillus reuteri* and found that the latter exhibited antibacterial activity against *Staphylococcus aureus* [34]. Cimmino et al. have isolated five DKPs, cyclo(L-Pro-L-Tyr), cyclo(L-Pro-L-Val), cyclo(D-Pro-D-Phe), cyclo(L-Pro-L-Leu), and cyclo(D-Pro-L-Tyr) with antibacterial activity against phytopathogenic Gram-positive bacterium *Rhodococcus fascians* LMG 3605 [35]. Cyclo(L-Pro-L-Phe), produced by *Pseudonocardia endophytica*, showed antibacterial activity against Gram-negative bacteria (*Xanthomonas campestris* and *Xanthomonas malvacearum*) and antifungal activity against (*Fusarium oxysporum* and *Fusarium solani*) [36]. Interestingly, mixing of cyclo(L-Phe-L-Pro) and cyclo(L-Leu-L-Pro) showed a good synergetic antimicrobial activity against *E. coli*, *Micrococcus luteus*, *S. aureus*, *C. albicans*, and *Cryptococcus neoformans* [37].

On the other hand, the MIC results for *A. alternata* 13A pure compound indicated that phragamide A (**7**) exhibited considerable antimicrobial activity against *P. aeruginosa*, *C. albicans*, and both Gram-positive strains (Table 3). Phragamide B (**8**) displayed moderate activity against *C. albicans* but showed weak activity against bacterial pathogens. Tenuazonic acid (**9**) exhibited a moderate antibacterial activity against Gram-positive strains, which is in accordance with a previous report of the antibacterial activity of **9** towards *Paenibacillus* larvae [38]. Altenusin (**11**) and alternariol (**12**) exhibited a similar antimicrobial activity towards *S. aureus*, *B. subtilis*, *P. aeruginosa*, and *C. albicans*. Altenusin (**11**) was reported to have antibacterial activity against *S. aureus* [39]. Moreover, Kjer et al. [40] reported strong antimicrobial activity of altenusin (**11**) against *S. aureus*, *P. aeruginosa*, and *C. albicans*. Additionally, alternariol monomethylether (**13**) and altertoxin I (**14**) showed weak antimicrobial activity against *S. aureus* and *C. albicans*, which is in agreement with previous reports by Sun et al. [25] who found that **13** had moderate antimicrobial activity against a panel of pathogenic bacteria and fungi including *S. aureus*, *Penicillium* sp., *Aspergillus* sp., and also weak antibacterial and antifungal properties observed for **12** and **13** [41]. Finally, altertoxin I (**14**), altertoxin II (**15**), and alterperylenol (**16**) exhibited a weak antibacterial activity against Gram-positive strains. Previously, **14** showed an antifungal activity against *Valsa ceratosperma*, a serious phytopathogenic fungus which causes canker disease for apples and induces the growth of lettuce seedlings [30], whereas **14** and **15** showed considerable inhibitory effect in an anti–HIV assay [29].

Our results show that out of the 16 compounds tested, only compounds **1** and **7** exhibited promising antimicrobial effects against Gram-positive strains. Moreover, all the other tested compounds showed a range of moderate, weak or no effect against the tested Gram-negative bacteria and *C. albicans*. This finding could be attributed to the loss (through isolation) of synergistic effects of the major and minor molecules present in the total fungal extract of *E. nigrum* M13 or *A. alternata* 13A which exhibited more promising results than the pure compounds alone, which is consistent with a previous study in the literature [37] and our own previous research [18].

2.3.2. Biofilm Inhibitory Activity

Bacterial biofilm formation has been found to play a critical role in the persistence of bacterial nosocomial infections. This phenomenon facilitates bacterial colonization on living or non-living surfaces and is associated with 65 to 80% of all clinical infections. Due to such adaptive changes, biofilm-forming bacteria are 10- to 1000-fold more resistant to conventional antibiotics, which thus presents a great challenge to develop antimicrobials specifically to treat biofilms [42]. Results of conventional antimicrobial susceptibility testing in vitro such as the MIC determination might not be appropriate to guide therapy for biofilm-associated infections. In fact, antimicrobial treatments based on MIC results often fail to eradicate surface-attached bacteria [43]. Consequently, we have screened the 16 pure isolated fungal metabolites against four clinical biofilm-forming pathogenic bacterial clinical isolates from Egyptian hospitals for their biofilm inhibitory activity using a microtiter biofilm plate assay. The obtained results showed that only three compounds **1**, **3**, and **5** from *E. nigrum* M13 displayed biofilm inhibitory activity against the tested clinical isolates. At 100 μg/mL, epicotripeptin (**1**) showed moderate biofilm inhibitory activity against the tested Gram-positive strains (55–70% inhibition), and weak activity against the tested Gram-negative strains (20–30%) (Figure 3). Cyclo(L-Pro-L-Ile) (**3**) and cyclo(L-Pro-L-Tyr) (**5**) exhibited a moderate biofilm formation inhibition against both Gram-positive strains but were not active against the tested Gram-negative strains. On the other hand, four pure compounds of *A. alternata* 13A were active towards the evaluated microorganisms at 100 μg/mL. Phragamides A and B (**7** and **8**), tenuazonic acid (**9**), and altechromone A (**10**) exhibited considerable biofilm formation inhibition against the tested Gram-positive strains (70–80% inhibition) and a moderate effect on Gram-negative strains (40–60%). Altenusin (**11**) exhibited moderate biofilm formation inhibition only against *B. subtilis* but a weak effect against the other three tested strains (Figure 3).

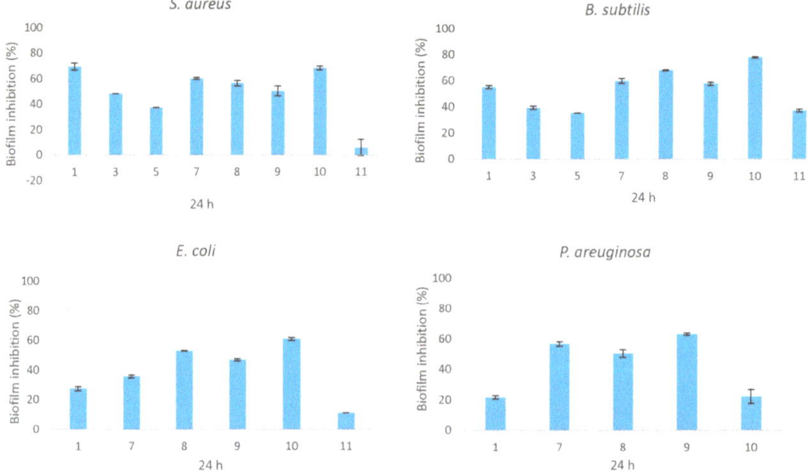

Figure 3. Biofilm inhibition effect of isolated pure compounds from *E. nigrum* M13 and *A. alternata* 13A against *S. aureus*, *B. subtilis*, *E. coli*, and *P. aeruginosa*. The biofilm was quantified using a microtiter plate and crystal violet assay. The bars on the graph represent the mean ± SD as a percentage of biofilm inhibition. Inactive compounds were not reported.

It is worth noting that cyclic dipeptides were previously reported to exhibit biofilm inhibitory effects against Gram-positive biofilm forming strains [44,45]. Additionally, tenuazonic acid (**9**) is known to have antibiofilm activity through interference with bacterial quorum sensing [46]. It should be noted that it is difficult to interpret whether the observed effects in our study are compounded by the impact of the compound on general cell viability, as non-selective (for example, membrane lytic) activity may well be involved. Further tests using red blood cells or other mammalian cells could provide a clearer picture but were beyond the focus of the present study.

3. Material and Methods

3.1. General Experimental Procedures

The structure characterisation of all compounds was based on ^{1}H NMR, ^{13}C NMR, COSY, HSQC, HMBC, and ROESY data, which were obtained on a Bruker Avance III 600 MHz spectrometer (BRUKER UK Ltd., Coventry, UK). An Agilent 1100 series HPLC system connected to the diode array G13158B detector was used for analytical and semi-preparative RP-HPLC purification (Agilent Technologies UK Ltd., Cheadle, UK). HPLC conditions were as follows: Phenomenex RP-C18 column (Luna 5 μ, 250 × 10 mm, L × i.d.) using a gradient of MeCN in H_2O over 35 min from 10 to 100% and ending up with 100% MeCN for 5 min at a flow rate of 1.5 mL/min. HRESIMS data were obtained using a Thermo LTQ Orbitrap coupled to an HPLC system (PDA detector, PDA autosampler, and pump). The following conditions were used: Capillary voltage of 45 V, capillary temperature of 260 °C, auxiliary gas flow rate of 10−20 arbitrary units, sheath gas flow rate of 40−50 arbitrary units, spray voltage of 4.5 kV, and mass range of 100−2000 amu (maximal resolution of 60,000). Optical rotations were recorded using a PerkinElmer 343 polarimeter (PerkinElmer, Waltham, MA, USA). UV spectra were obtained using a PerkinElmer Lambda2 UV/Vis spectrometer (PerkinElmer, Waltham, MA, USA).

3.2. Fungal Strains

The fungal strain M13 was obtained from *Thalassia hemprichii* leaves collected from Makady bay, South Hurghada, Red sea, Egypt and identified as *Epicoccum nigrum* based on its morphological features, together with ITS rDNA (GenBank accession number MK953943) and β-tubulin (GenBank accession number MT184348) phylogenetic analysis [18]. Strain 13A was isolated from the leaves of *Phragmites australis* collected from Lake El-Bida, Wadi El-Natrun depression, Beheira Governorate, Egypt, and identified as *Alternaria alternata* based on its morphological characteristics and its ITS rDNA sequence (GenBank accession number MK248606) [18].

3.3. Cultivation and Fermentation of Endophytic Fungi

The strain M13 was grown on a modified marine ISP2 broth medium (consisting of yeast extract 0.4%, malt extract 0.4%, and dextrose 0.4% for 1 L of distilled water incorporated with sea salts (consisting of 0.025% KI and $MgSO_4$, 0.05% $CaCl_2$ and 0.5% NaCl) adjusted to pH 6 before sterilization. Equal-sized agar plugs with mycelium were aseptically transferred to broth media (12 L) in 1 L conical flasks and incubated for 28 days under static conditions.

For the strain 13A, a seed culture was prepared by inoculation of the pure fungal mycelia into a 250 mL Erlenmeyer flask containing 100 mL of potato dextrose-SW broth (PDB-SW) medium and incubated at 28 °C for 4 days. In addition, 5 mL of the broth culture were transferred into 1 L conical flasks (10 L) containing a rice medium: 100 g commercial rice and 100 mL 50% seawater. The flasks were incubated for 15 days at 28 °C.

3.4. Extraction and Isolation

For the M13 isolate, the resulting thick mycelium bed was separated and extracted with MeOH at the end of the fermentation period. Then, about 5 g of Diaion HP20 was added to each flask containing 500 mL of broth media and shaked at 180 rpm for 5 h. Thereafter, Diaion HP20 was filtered and extracted with MeOH (3 × 500 mL). Both MeOH extracts were combined and evaporated under a vacuum to obtain the total crude extract, which was defatted with *n*-hexane and subsequently fractionated with DCM and EtOAc. Then, DCM and EtOAc were analysed for their metabolite profile via ^{1}H NMR spectroscopy and analytical HPLC (Agilent Technologies UK Ltd., Cheadle, UK, Phenomenex RP-C18 column Luna 5 μm, 250 × 4.60 mm, L × i.d.) at a flow rate of 1.0 mL/min using a gradient of 10–100% of MeCN in H_2O over a period of 25 min and 100% MeCN for 5 min. Further purification and isolation of compounds from the DCM fraction were carried out using semi-preparative RP-HPLC (Phenomenex RP-C18 column Luna 5 μm, 150 × 10 mm, L × i.d.) using a gradient of 10 to 100% MeCN in H_2O over a period of 35 min followed by 100% MeCN for 5 min at a flow rate of 1.5 mL/min to produce compounds **1** (R_t 16.5 min, 2.0 mg), **2** (R_t 10.9 min, 2.1 mg), **3** (R_t 14.0 min, 3.6 mg), **4** (R_t 15.8 min, 4.5 mg), **5** (R_t 11.1 min, 2.0 mg), and **6** (R_t 18.4 min, 2.5 mg).

For the 13A isolate, after the incubation period, the fermented rice was soaked overnight in EtOAc (1:1 *v/v*), and the extract was then collected and evaporated. The obtained crude extract was suspended in 50% aqueous methanol (100 mL) and sequentially partitioned using *n*-hexane, DCM, and EtOAc. Purification of DCM and EtOAc fractions over different chromatographic techniques were carried out. Final purification of DCM extract using semi preparative HPLC over a period of 60 min yielded compounds **7** (R_t

14.6 min, 2.0 mg), **8** (R_t 25.2 min, 1.1 mg), **9** (R_t 19.1 min, 37 mg), **10** (R_t 21.6 min, 1.7 mg), **12** (R_t 28.9 min, 4 mg), **13** (R_t 38.6 min, 4.2 mg), while the EtOAc fraction over a period of 25 min following the same HPLC conditions as above furnished **11** (R_t 9.2 min, 2 mg), **14** (R_t 14.3 min, 2.0 mg), **15** (R_t 11.6 min, 1.4 mg), and **16** (R_t 18.2 min, 2.5 mg).

Epicotripeptin (**1**): Pale yellow solid; $[\alpha]_D^{25} = -19.65$ (c 0.01, MeOH); UV (MeOH) λ_{max} (logε) 218 (3.90), 252 (3.15), 284 (2.75) nm; ^{1}H NMR and ^{13}C NMR data, see Table 1; HRESIMS [M + Na]$^+$ at m/z 453.1899 $C_{25}H_{26}N_4O_3Na$ (Calc. 453.1897).

Phragamide A (**7**): Pale yellow amorphous solid; $[\alpha]_D^{25} = -13.7$ (c=0.3, MeOH), λ_{max}^{MeOH}: 214, 224, 254 nm; ^{1}H NMR and ^{13}C NMR data, see Table 2; HRESIMS: [M - H]$^-$ at m/z 214.1082 $C_{10}H_{17}NO_4$ (Calc. 214.1085).

Phragamide B (**8**): Brown oil; $[\alpha]_D^{25} = -8.3$ (c=0.2, MeOH), λ_{max}^{MeOH}: 214, 224, 256 nm; ^{1}H NMR and ^{13}C NMR data, see Table 2; HRESIMS: [M - H]$^-$ at m/z 184.0979 $C_9H_{15}NO_3$ (Calc. 184.0979).

3.5. Advanced Marfey's Analysis

Epicotripeptin (**1**, 0.2 mg) or phragamide A (**7**, 0.2 mg) was hydrolysed in 0.4 mL of 6 N HCl at 110 °C for 24 h. The reaction mixture was subsequently cooled, the solvent evaporated under N_2, and residual HCl completely removed by freeze-drying the sample overnight. The hydrolysate was dissolved in 50 µL of H_2O. To 50 µL of a 50 mM standard amino acid solution (aqueous) or the hydrolysate, 20 µL of 1 M NaHCO$_3$ and 100 µL of 1% FDLA (1-fluoro-2,4-dinitrophenyl-5-L-leucinamide) in acetone were added. The reaction mixtures were incubated at 40 °C for 1 h, with frequent mixing. After the mixtures had been cooled at room temperature, the reactions were quenched by the addition of 10 µL of 2 N HCl. The samples were diluted by increasing the volume to 1 mL with MeOH. A 5 µL aliquot of each sample was analysed by LC-MS using a Kinetex C18 column (2.6 µm, 100 Å, 2.10 × 100 mm), eluted with a solvent gradient of 10 to 80% MeCN in H_2O containing 0.1% formic acid over 30 min, at a flow rate of 1 mL/min. The comparison of the retention times of derivatised hydrolysed amino acids with that of derivatised D- and L-amino acid standards revealed the L-configuration for the amino acid residues [19].

3.6. Mosher Ester Analysis

To 0.5 mg of compound **7**, 180 µL of dry pyridine-d_5 was added, and the solution was transferred into an NMR tube. To initiate the reaction, 20 µL of (R)-MTPA-Cl was added with careful shaking and then monitored immediately by ^{1}H NMR after 30 min. Analogously, for an equal amount of **7** dissolved in 180 µL of dry pyridine-d_5, a second NMR tube with 20 µL of (S)-MTPA-Cl was reacted for 30 min, to afford the mono (R)-MTPA and (S)-MTPA ester derivatives of **7**, respectively [24]. The chemical shift difference ($\Delta\delta = \delta_{S\text{-MTPA-ester}} - \delta_{R\text{-MTPA-ester}}$) of the protons near C-7 were observed.

3.7. Bioactivity Studies

3.7.1. Antimicrobial Assay

For the antimicrobial testing, two Gram-positive bacteria (*Bacillus subtilis* ATCC6633 and *Staphylococcus aureus* NRRLB-767), five Gram-negative bacteria (*Escherichia coli* ATCC25955, *Pseudomonas aeruginosa* ATCC10145, *Klebsiella pneumonia* ATCCBAA-1705, and *Proteus vulgaris* ATTC7829), and one yeast (*Candida albicans* ATCC10231) were obtained from the Microbiology and Immunology Department, Faculty of Medicine, Al-Azhar University, Egypt. The antimicrobial assay and MIC were performed as described by Ingebrigtsen et al. 2016 [47], in which 10 µL of bacterial or fungal suspension at the log phase were added to 180 µL of lysogeny broth (LB), followed by the addition of 10 µL of the tested pure compounds. The final concentrations of the mixture were 50, 40, 30, 20, 10, 5, 2.5, 1.25, 0.62, 0.31, and 0.15 µg/mL. After inoculation, the plates were incubated overnight at 37 °C for 24 h. Then, the absorbance was measured at 600 nm for bacteria and 340 nm for *C. albicans* using a Spectrostar Nano Microplate Reader (BMG LABTECH GmbH, Allmendgrun, Germany). The MIC were reported as the average of the lowest concentrations with no observable growth of microorganisms. The MIC were determined in two independent experiments. Ciprofloxacin and nystatin were used as positive controls.

3.7.2. Biofilm Inhibitory Activity

The antibiofilm formation activity was performed using a microtiter plate assay. The effect of the fungal 13A and M13 crude extract and the isolated pure compounds on the biofilm formation on four biofilm-forming pathogenic bacterial clinical isolates from Egyptian hospitals *(B. subtilis, s. aureus, p. aeruginosa,* and *E. coli)* was measured in a 96-well polystyrene microtiter plate [43]. The test bacteria were first inoculated in a 100 mL Erlenmeyer flask containing LB media and incubated at 37 °C overnight in an orbital shaker at 150 rpm. Each individual well was filled with 180 µL of LB broth, then inoculated with 10 µL of pathogenic bacterial suspension and incubated for 12 h. To this, 10 µL of the test crude and pure compounds was added along with the control (without the test sample) and incubated statically at 37 °C for 24 h. After incubation, the contents of the wells were carefully removed and washed with 200 µL per well of phosphate buffer saline (PBS) pH 7.2, to remove the free-floating bacteria. The microplate was air-dried for 1 h, stained with 200 µL/well of crystal violet solution (0.1%, *w/v*), and left at room temperature for 10 min. The microplate wells were washed three times with 200 µL/well distilled water and kept for drying to remove the excess stain. To solubilise the dye, the dried microplate was washed with 200 µL/well of 95% ethanol, and the intensity was measured at an optical density of 570 nm using a Spectrostar Nano Microplate Reader (Spectrostar Nano, Belfast, Northern Ireland).

Supplementary Materials: The following are available online at https://www.mdpi.com/article/10.3390/md19040232/s1, Table S1: Chemical shift difference of (S)–MTPA and (R)–MTPA ester of 6 (pyridine-d_5 at 298 K, Figures S1–S30: 1D and 2D NMR spectra of compound **1, 7,** and **8** and ^{1}H NMR of all the known compounds.

Author Contributions: Conceptualisation, A.A.H., R.E., and M.E.R.; methodology, M.M.Q., A.A.H., A.S.I.H., S.S., M.A., and M.E.E.; formal analysis, M.M.Q., A.A.H., S.S., A.S.I.H., L.B., M.A., M.E.E., M.E.R., and R.E.; investigation, M.M.Q., A.A.H., S.S., and L.B.; data curation, M.M.Q., A.A.H., S.S., M.A., M.E.E., L.B., M.E.R., and R.E.; writing—original draft preparation, M.M.Q., A.A.H., A.S.I.H., M.A., M.E.E., and S.S.; writing and editing, A.A.H., M.E.R., and R.E.; funding acquisition, A.A.H., M.E.R., and R.E. All authors have read and agreed to the published version of the manuscript.

Funding: This research was funded by the British Council Newton Fund Institutional Links Project number 261781172 and The Science and Technology Development Fund (STDF) project number 27701.

Institutional Review Board Statement: Not applicable.

Informed Consent Statement: Not applicable.

Acknowledgments: We would like to thank the College of Physical Sciences, University of Aberdeen, for the provision of infrastructure and facilities in the Marine Biodiscovery Centre.

Conflicts of Interest: The authors declare no conflict of interest.

References

1. Nogawa, T.; Kawatani, M.; Okano, A.; Futamura, Y.; Aono, H.; Shimizu, T.; Kato, N.; Kikuchi, H.; Osada, H. Structure and biological activity of Metarhizin C, a stereoisomer of BR-050 from *Tolypocladium album* RK17-F0007. *J. Antibiot.* **2019**, *72*, 996–1000. [CrossRef] [PubMed]

2. Bernan, V.S.; Greenstein, M.; Maiese, W.M. Marine microorganisms as a source of new natural products. *Adv. Appl. Microbiol.* **1997**, *43*, 57–90. [PubMed]

3. El-Hady, F.K.A.; Shaker, K.H.; Souleman, A.M.; Fayad, W.; Abdel-Aziz, M.S.; Hamed, A.A.; Iodice, C.; Tommonaro, G. Comparative correlation between chemical composition and cytotoxic potential of the coral-associated fungus *Aspergillus* sp. 2C1-EGY against human colon cancer cells. *Curr. Microbiol.* **2017**, *74*, 1294–1300. [CrossRef] [PubMed]

4. El-Neekety, A.A.; Abdel-Aziz, M.S.; Hathout, A.S.; Hamed, A.A.; Sabry, B.A.; Ghareeb, M.A.; Aly, S.E.; Abdel-Wahhab, M.A. Molecular identification of newly isolated non-toxigenic fungal strains having antiaflatoxigenic, antimicrobial and antioxidant activities. *Der Pharm. Chem.* **2016**, *8*, 121–134.

5. Blunt, J.W.; Copp, B.R.; Munro, M.H.; Northcote, P.T.; Prinsep, M.R. Marine natural products. *Nat. Prod. Rep.* **2011**, *28*, 196–268. [CrossRef]

6. Imhoff, J. Natural products from marine fungi—Still an underrepresented resource. *Mar. Drugs* **2016**, *14*, 19. [CrossRef] [PubMed]

7. Schol–Schwarz, M.B. The genus *Epicoccum* Link. *Trans. Brit. Mycol. Soc.* **1959**, *42*, 149–173. [CrossRef]

8. Meena, M.; Swapnil, P.; Upadhyay, R.S. Isolation, characterisation, and toxicological potential of *Alternaria*-mycotoxins (TeA, AOH and AME) in different *Alternaria* species from various regions of India. *Sci. Rep.* **2017**, *7*. [CrossRef] [PubMed]

9. Baute, M.A.; Deffieux, G.; Baute, R.; Neveu, A. New antibiotics from the fungus *Epicoccum nigrum*. *J. Antibiot.* **1978**, *31*, 1099–1101. [CrossRef] [PubMed]

10. Lou, J.; Fu, L.; Peng, Y.; Zhou, L. Metabolites from *Alternaria* fungi and their bioactivities. *Molecules* **2013**, *18*, 5891–5935. [CrossRef] [PubMed]

11. Thomma, B.P.H.J. *Alternaria* spp.: From general saprophyte to specific parasite. *Mol. Plant Pathol.* **2003**, *4*, 225–236. [CrossRef] [PubMed]

12. de Lima Favaro, L.C.; de Souza Sebastianes, F.L.; Araújo, W.L. *Epicoccum nigrum* P16, a sugarcane endophyte, produces antifungal compounds and induces root growth. *PLoS ONE* **2012**, *7*, e36826.

13. Bamford, P.C.; Norris, G.L.F.; Ward, G. Flavipin production by *Epicoccum* spp. *Trans. Brit. Mycol. Soc.* **1961**, *44*, 354–356. [CrossRef]

14. Brown, A.E.; Finlay, R.; Ward, J.S. Antifungal compounds produced by *Epicoccum purpurascens* against soil born plant pathogenic fungi. *Soil. Biol. Biochem.* **1987**, *19*, 657–664. [CrossRef]

15. Guo, H.; Sun, B.; Gao, H.; Chen, X.; Liu, S.; Yao, X.; Liu, X.; Che, Y. Diketopiperazines from the *Cordyceps*– colonising fungus *Epicoccum nigrum*. *J. Nat. Prod.* **2009**, *72*, 2115–2119. [CrossRef]

16. Wang, J.-M.; Ding, G.-Z.; Fang, L.; Dai, J.-G.; Yu, S.-S.; Wang, Y.-H.; Chen, X.-G.; Ma, S.-G.; Qu, J.; Xu, S.; et al. Thiodiketopiperazines produced by endophytic fungus *Epicoccum nigrum*. *J. Nat. Prod.* **2010**, *73*, 1240–1249. [CrossRef]

17. Sun, H.-H.; Mao, W.-J.; Jiao, J.-Y.; Xu, J.-C.; Li, H.-Y.; Chen, Y.; Qi, X.-H.; Chen, Y.-L.; Xu, J.; Zhao, C.-Q.; et al. Structural characterisation of extracellular polysaccharides produced by the marine fungus *Epicoccum nigrum* JJY–40 and their antioxidant activities. *Mar. Biotechnol.* **2011**, *13*, 1048–1055. [CrossRef]

18. Hamed, A.A.; Soldatou, S.; Qader, M.M.; Arjunan, S.; Miranda, K.J.; Casolari, F.; Pavesi, C.; Diyaolu, O.A.; Thissera, B.; Eshelli, M.; et al. Screening fungal endophytes derived from under-explored Egyptian marine habitats for antimicrobial and antioxidant properties in factionalised textiles. *Microorganisms* **2020**, *8*, 1617. [CrossRef]

19. Fujii, K.; Ikai, Y.; Oka, H.; Suzuki, M.; Harada, K. A nonempirical method using LC/MS for determination of the absolute configuration of constituent amino acids in a peptide: combination of Marfey's method with mass spectrometry and its practical application. *Anal. Chem.* **1997**, *69*, 5146–5151. [CrossRef]

20. He, R.; Wang, B.; Wakimoto, T.; Wang, M.; Zhu, L.; Abe, I. Cyclodipeptides from metagenomic library of a Japanese marine sponge. *J. Braz. Chem. Soc.* **2013**, *24*, 1926–1932. [CrossRef]

21. Cimmino, A.; Puopolo, G.; Perazzolli, M.; Andolfi, A.; Melck, D.; Pertot, I.; Evidente, A. Cyclo(L–Pro–L–Tyr), The fungicide isolated from *lysobacter capsici* az78: A structure–activity relationship study. *Chem. Heterocycl. Comp.* **2014**, *50*, 290–295. [CrossRef]

22. Daoud, N.N.; Foster, H.A. Antifungal activity of *Myxococcus* species 1 production, physiochemical and biological properties of antibiotics from *Myxococcus flavus* S110 (Myxobacterales). *Microbios* **1993**, *73*, 173–184.

23. Hoye, T.R.; Jeffrey, C.S.; Shao, F. Mosher ester analysis for the determination of absolute configuration of stereogenic (chiral) carbinol carbons. *Nat. Protoc.* **2007**, *2*, 2451–2458. [CrossRef]

24. Smelcerovic, A.; Yancheva, D.; Cherneva, E.; Petronijevic, Z.; Lamshoeft, M.; Herebian, D. Identification and synthesis of three cyclodidepsipeptides as potential precursor of enniatin B in *Fusarium sporotrichioides*. *J. Mol. Structure* **2011**, *985*, 397–402. [CrossRef]

25. Sun, J.; Awakawa, T.; Noguchi, H.; Abe, I. Induced production of mycotoxins is an endophytic fungus from the medicinal plant *Datura stramonium* L. *Bioorg. Med. Chem. Lett.* **2012**, *22*, 6397–6400. [CrossRef]

26. Hu, L.; Chen, N.; Hu, Q.; Yang, C.; Yang, Q.; Qang, F.-F. An unusual piceatannol dimer from *Rheum austral* D. Don with antioxidant activity. *Molecules* **2014**, *19*, 11453–11464. [CrossRef]

27. Kamisuki, S.; Takahashi, S.; Mizushina, Y.; Hanashima, S.; Kuramochi, K.; Kobayashi, S.; Sakaguchi, K.; Nakata, T.; Sugawara, F. Total synthesis of dehydroaltenusin. *Tetrahedron* **2004**, *60*, 5695–5700. [CrossRef]

28. Mousa, W.K.; Schwan, A.; Davidson, J.; Strange, P.; Liu, H.; Zhou, T.; Auzanneau, F.I.; Raizada, M.N. An endophytic fungus isolated from finger millet (*Eleusine coracana*) produces anti–fungal natural products. *Front. Microbiol.* **2015**, *6*. [CrossRef]

29. Bashyal, B.P.; Wellensiek, B.P.; Ramakrishnan, R.; Faeth, S.H.; Ahmed, N.; Gunatilaka, A.A.L. Aflatoxins with potent anti–HIV activity from *Alternaria tenuissima* QUE1Se, a fungal endophyte of *Quercus emoryi*. *Bioorg. Med. Chem.* **2014**, *22*, 6112–6116. [CrossRef]

30. Okuno, T.; Natsume, I.; Sawai, K.; Sawamura, K.; Furusaki, A.; Matsumoto, T. Structure of antifungal and phytotoxic pigments produced by *Alternaria* sp. *Tetrahedron Lett.* **1983**, *24*, 5653–5656. [CrossRef]

31. O'Meara, S. Antimicrobial resistance. *Nature* **2020**, *586*, S49. [CrossRef]

32. Fdhila, F.; Vazquez, V.; Sanchez, J.L.; Riguera, R. DD-diketopiperazines: Antibiotics active against *Vibrio anguillarum* isolated from marine bacteria associated with cultures of *Pecten maximus*. *J. Nat. Prod.* **2003**, *66*, 1299–1301. [CrossRef]

33. Houston, D.R.; Synstad, B.; Eijsink, V.G.; Stark, M.J.; Eggleston, I.M.; van Aalten, D.M. Structure-based exploration of cyclic dipeptide chitinase inhibitors. *J. Med. Chem.* **2004**, *47*, 5713–5720. [CrossRef]

34. Li, J.; Wangb, W.; Xua, S.X.; Magarveyb, N.A.; McCormicka, J.K. *Lactobacillus reuteri*-produced cyclic dipeptides quench agr-mediated expression of toxic shock syndrome toxin-1 in staphylococci. *Proc. Natl. Acad. Sci. USA* **2011**, *108*, 3360–3365. [CrossRef]

35. Cimmino, A.; Bejarano, A.; Masi, M.; Puopolo, G.; Evidente, A. Isolation of 2,5-diketopiperazines from *Lysobacter capsici* AZ78 with activity against *Rhodococcus fascians*. *Nat. Prod. Res.* **2020**, *30*. [CrossRef]

36. Mangamuri, U.K.; Muvva, V.; Poda, S.; Manavathi, B.; Bhujangarao, C.; Yenamandra, V. Chemical characterization and bioactivity of diketopiperazine derivatives from the mangrove derived *Pseudocardia endophytica*. *Egypt. J. Aquat. Res.* **2016**, *42*, 169–175. [CrossRef]

37. Rhee, K.-H. Cyclic dipeptides exhibit synergistic, broad spectrum antimicrobial effects and have anti-mutagenic properties. *Int. J. Antimicrob. Agents* **2004**, *24*, 423–427. [CrossRef]

38. Gallardo, G.L.; Peña, N.I.; Chacana, P.; Terzolo, H.R.; Cabrera, G.M. L-Tenuazonic acid, a new inhibitor of *Paenibacillus* larva. *World J. Microbiol. Biotechnol.* **2004**, *20*, 609–612. [CrossRef]

39. Xu, X.; Zhao, S.; Wei, J.; Fang, N.; Yin, L.; Sun, J. Porric acid D from marine-derived fungus *Alternaria* sp. isolated from Bohai sea. *Chem. Nat. Compd.* **2012**, *47*, 893–895. [CrossRef]

40. Kjer, J.; Wray, V.; Edrada-Ebel, R.A.; Ebel, R.; Pretsch, A.; Lin, W.H.; Proksch, P. Xanalteric acids I and II and related phenolic compounds from an endophytic *Alternaria* sp. isolated from the mangrove plant *Sonneratia alba*. *J. Nat. Prod.* **2009**, *72*, 2053–2057. [CrossRef]

41. Gu, W. Bioactive metabolites from *Alternaria brassicicola* ML–P08, an endophytic fungus residing in *Malus halliana*. *World J. Microbiol. Biotechnol.* **2009**, *25*, 1677–1683. [CrossRef]

42. Pletzer, D.; Hancock, R.E.W. Antibiofilm peptides: Potential as broad-spectrum agents. *J. Bacteriol.* **2016**, *198*, 2572–2578. [CrossRef]

43. Antunes, A.L.; Trentin, D.S.; Bonfanti, J.W.; Pinto, C.C.; Perez, L.R.; Macedo, A.J.; Barth, A.L. Application of a feasible method for determination of biofilm antimicrobial susceptibility in staphylococci. *APMIS* **2010**, *118*, 873–877. [CrossRef]

44. Yu, X.; Li, L.; Sun, S.; Chang, A.; Dai, X.; Li, H.; Wang, Y.; and Hu Zhu, H. A cyclic dipeptide from marine fungus *Penicillium chrysogenum* DXY-1 exhibits anti-quorum sensing activity. *ACS Omega* **2021**, *6*, 7693–7700. [CrossRef] [PubMed]

45. Simon, G.; Bérubé, C.; Voyer, N.; Grenier, D. Anti-biofilm and anti-adherence properties of novel cyclic dipeptides against oral pathogens. *Bioorg. Med. Chem.* **2019**, *27*, 2323–2331. [CrossRef] [PubMed]

46. Dobretsov, S.; Teplitski, M.; Bayer, M.; Gunasekera, S.; Proksch, P.; Paul, V.J. Inhibition of marine biofouling by bacterial quorum sensing inhibitors. *Biofouling* **2011**, *27*, 893–905. [CrossRef] [PubMed]

47. Ingebrigtsen, R.A.; Hansen, E.; Andersen, J.H.; Eilertsen, H.C. Light and temperature effects on bioactivity in diatoms. *J. Appl. Phycol.* **2016**, *28*, 939–950. [CrossRef]

Article

Bio-Guided Isolation of Antimalarial Metabolites from the Coculture of Two Red Sea Sponge-Derived *Actinokineospora* and *Rhodococcus* spp.

Hani A. Alhadrami [1,2,†], Bathini Thissera [3,†], Marwa H. A. Hassan [4], Fathy A. Behery [5,6], Che Julius Ngwa [7], Hossam M. Hassan [4,8], Gabriele Pradel [7], Usama Ramadan Abdelmohsen [9,10,*] and Mostafa E. Rateb [3,*]

1 Department of Medical Laboratory Technology, Faculty of Applied Medical Sciences, King Abdulaziz University, P.O. BOX 80402, Jeddah 21589, Saudi Arabia; hanialhadrami@kau.edu.sa
2 Molecular Diagnostic Laboratory, King Abdulaziz University Hospital, King Abdulaziz University, P.O. BOX 80402, Jeddah 21589, Saudi Arabia
3 School of Computing, Engineering & Physical Sciences, University of the West of Scotland, Paisley PA1 2BE, UK; bathini.thissera@uws.ac.uk
4 Department of Pharmacognosy, Faculty of Pharmacy, Beni-Suef University, Beni-Suef 62514, Egypt; mh_elsefy@yahoo.com (M.H.A.H.); abuh20050@yahoo.com (H.M.H.)
5 Department of Pharmacognosy, Faculty of Pharmacy, Mansoura University, Mansoura 35516, Egypt; fathy.behery@riyadh.edu.sa
6 Department of Pharmacy, College of Pharmacy, Riyadh Elm University, Riyadh 11681, Saudi Arabia
7 Division of Cellular and Applied Infection Biology, Institute of Zoology, RWTH Aachen University, 52074 Aachen, Germany; ngwa.che@bio2.rwth-aachen.de (C.J.N.); pradel@bio2.rwth-aachen.de (G.P.)
8 Department of Pharmacognosy, Faculty of Pharmacy, Nahda University, Beni-Suef 62514, Egypt
9 Department of Pharmacognosy, Faculty of Pharmacy, Minia University, Minia 61519, Egypt
10 Department of Pharmacognosy, Faculty of Pharmacy, Deraya University, New Minia 61111, Egypt
* Correspondence: usama.ramadan@mu.edu.eg (U.R.A.); Mostafa.Rateb@uws.ac.uk (M.E.R.)
† These authors are equally contributed.

Citation: Alhadrami, H.A.; Thissera, B.; Hassan, M.H.A.; Behery, F.A.; Ngwa, C.J.; Hassan, H.M.; Pradel, G.; Abdelmohsen, U.R.; Rateb, M.E. Bio-Guided Isolation of Antimalarial Metabolites from the Coculture of Two Red Sea Sponge-Derived *Actinokineospora* and *Rhodococcus* spp. *Mar. Drugs* **2021**, *19*, 109. https://doi.org/10.3390/md19020109

Academic Editor: Marialuisa Menna

Received: 18 January 2021
Accepted: 9 February 2021
Published: 12 February 2021

Publisher's Note: MDPI stays neutral with regard to jurisdictional claims in published maps and institutional affiliations.

Abstract: Coculture is a productive technique to trigger microbes' biosynthetic capacity by mimicking the natural habitats' features principally by competition for food and space and interspecies crosstalks. Mixed cultivation of two Red Sea-derived actinobacteria, *Actinokineospora spheciospongiae* strain EG49 and *Rhodococcus* sp. UR59, resulted in the induction of several non-traced metabolites in their axenic cultures, which were detected using LC–HRMS metabolomics analysis. Antimalarial guided isolation of the cocultured fermentation led to the isolation of the angucyclines actinosporins E (**1**), H (**2**), G (**3**), tetragulol (**5**) and the anthraquinone capillasterquinone B (**6**), which were not reported under axenic conditions. Interestingly, actinosporins were previously induced when the axenic culture of the *Actinokineospora spheciospongiae* strain EG49 was treated with signalling molecule *N*-acetyl-D-glucosamine (GluNAc); this finding confirmed the effectiveness of coculture in the discovery of microbial metabolites yet to be discovered in the axenic fermentation with the potential that could be comparable to adding chemical signalling molecules in the fermentation flask. The isolated angucycline and anthraquinone compounds exhibited in vitro antimalarial activity and good biding affinity against lysyl-tRNA synthetase (PfKRS1), highlighting their potential developability as new antimalarial structural motif.

Keywords: *Actinokineospora*; *Rhodococcus*; co-culture; metabolomics; antimalarial; docking

1. Introduction

Exploring microbial forms of communication and utilising them in the production of secondary metabolites is of benefit in the process of natural products drug discovery [1]. Thus far, microbial secondary metabolites remain the major source for antimicrobial agents [2–4]. However, gene sequencing of many microbial genome showed that several species, mainly filamentous bacteria and fungi, apply a considerable part of their genes for secondary metabolism (10–15%) [5,6]. Remarkably, most microorganism genes are silent and have no role during laboratory cultivation [7]. Robert Koch used cultures for only one species of microorganism "axenic growth" to provide an apparent elucidation for this phenomenon of silent genes [8]. Microorganism culture in laboratory included macro- and micro-nutrients, constant temperature, adjusted pH, high water activity, and no contact with other world microbes, and thus a significant part of microorganisms' secondary metabolites, mainly those responsible for interaction, communication, or involving in fights with other species, are not found in microbial metabolites. Therefore, the new approach of co-cultivation provides a massive chance to motivate the silent genes and increase the opportunity to discover cryptic bioactive metabolites [1]. The fortune of "uncultivable" diversity is represented as the "microbial dark matter", the part of microorganisms that was unable to be cultivated in the laboratory until now [1]. The first reported mixed culture was in 1918; it was a coculture of *Escherichia coli* and *Bacillus paratyphosus* [9]. Up until now, natural product discovery, biotechnology, and microbiology scientists work on the discovery of coculture or mixed culture experiments to study the difference in the secondary metabolites produced during these trials compared to "axenic growth" [1]. From the examples for coculture and secondary metabolites production, mixed culture of *Acremonium* sp. and *Mycogonerosea* that produced new lipoaminopeptides, the acremostatins A-C [10]. Coculture of the marine-derived fungi *Aspergillus fumigatus* together with two desert bacterial isolates yielded new compounds, namely, luteoride D and pseurotin G [11]. Furthermore, a new N-methoxypyridone was discovered from a mixed fermentation of two endophytic fungi *Camporesia sambuci* and *Epicoccum sorghinum* isolated from the fruit of *Rhodomyrtus tomentosa* plant, collected on the Big Island in Hawaii [12]. New antifungal pulicatin derivatives H and I were induced following coculturing of plant-derived bacterium *Pantoea agglomerans* and the fungus *Penicillium citrinum* [13]. All these examples exemplify that co-cultivation of microorganisms induces new secondary metabolites that can be recommended as an appropriate way to produce diverse bioactive microbial metabolites.

Malaria was identified as a lethal disease caused by *Plasmodium* parasites, which infect humans through the malaria vector Anopheles mosquitoes. To date, five species of parasites have been identified as causatives of malaria in humans; two of them cause serious infections—*P. falciparum* and *P. vivax*. Studying the malaria cases worldwide revealed that 29 countries accounted for 95% of malaria cases. The majority of cases (82%) and deaths (94%) were reported in the WHO African region, followed by the WHO South-East Asia region (10% cases and 3% deaths) (https://www.who.int/publications/i/item/9789240015791, accessed on 30 January 2021). Malaria management and suppression require a complicated method. Up until now, two important antimalarial drugs are used to control infection. These two bitter principle drugs are derived from plants: artemisinin obtained from *Artemisia annua* L. (4th century, China), and quinine alkaloid obtained from *Cinchona* sp. (17th century, South America) [14]. The WHO recommends artemisinin combination therapy (ACT) as the first treatment plan in most malarial cases. However, in 2009, resistance to artemisinin combination therapy was reported. The emerging of drug resistance led to increased malaria cases and an increase in mortality [15]. Thus, the WHO endorsed using a combination of two drugs that work in different mechanisms to control drug resistance. The latest reports from Southeast Asia and India [16] showed the limitation of disease resistance to combination of artemisinin and other drugs as mefloquine and piperaquine [17]. Lacking effective new generation of medicines against malarial invasion, the number of new cases and deaths may rise. Thus, developing antimalarial therapeutics is important to save a large number of lives.

In this work, we discuss the application of co-cultivation of two actinobacteria: *Actinokineospora spheciospongiae* strain EG49 and *Rhodococcus* sp. UR59 recovered from Red Sea sponges as a strategy to stimulate silent genes and discover cryptic secondary metabolites within both strains. Additionally, antimalarial-guided fractionation of the bacterial coculture extract led to the isolation and characterisation of a few active metabolites against *P. falciparum*. A potential antimalarial target is proposed on the basis of molecular docking experiments against a number of reported targets.

2. Results and Discussions

2.1. Identification of Red Sea Sponge-Associated Actinobacteria

Two Red sea sponge-associated actinobacteria were isolated and taxonomically identified. *Actinokineospora spheciospongiae* strain EG49 was previously characterised [18,19]. The other actinobacterial strain was taxonomically identified as *Rhodococcus* sp. UR59, according to its morphology and its 16S rRNA genome sequence and phylogenetic analyses (Figure 1).

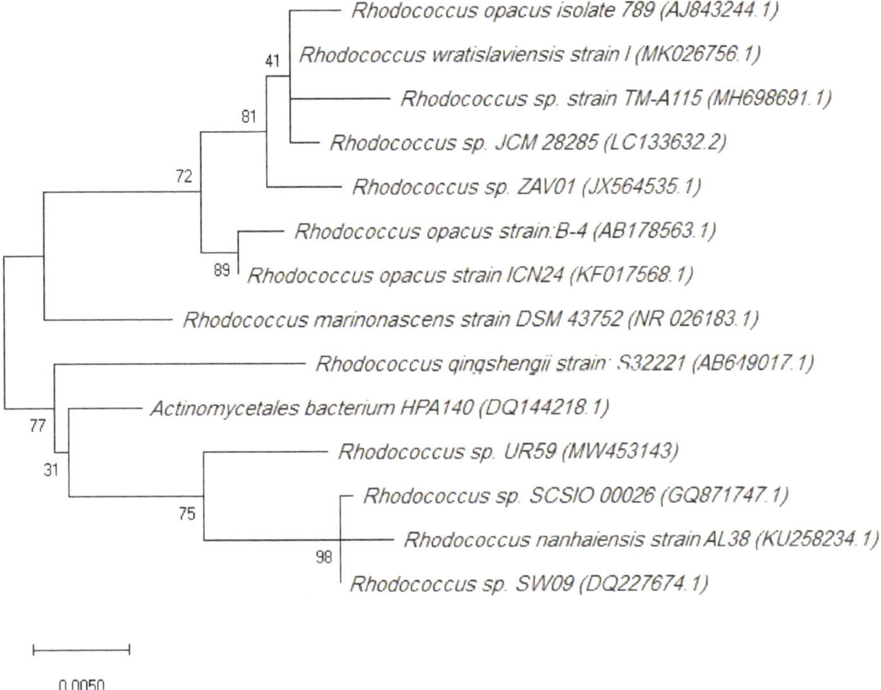

Figure 1. Phylogenetic tree of the *Rhodococcus* sp. UR59 isolate and the closest relatives in terms of the 16S rRNA gene marker. The accession numbers are indicated in brackets.

2.2. Metabolomics Analysis of the Coculture Extract of Actinokineospora spheciospongiae Strain EG49 and Rhodococcus sp. UR59 Using LC–HRMS

The analysis of the metabolomics data (Table 1) revealed 34 microbial secondary metabolites, of which 9 were detected from *Actinokineospora spheciospongiae* strain EG49 and the rest from *Rhodococcus* sp. UR59. Additionally, the analysis revealed the presence of diverse microbial chemical classes, namely, 10 angucyclines, 7 peptides, 3 macrolides, 3 anthraquinones, 2 polyenes, 2 polyethers, 2 phenolics, and 1 glycolipid. The predicted formula $C_{16}H_{18}N_2O_4$ was annotated as mitomycin-K [20,21], whereas $C_{18}H_{14}O_6$ was dereplicated as fluostatin-B, an inhibitor of dipeptidyl peptidase III that was previously isolated from *Streptomyces* sp. TA-3391 [22]. Moreover, the predicted formulas $C_{32}H_{33}O_{15}$ and $C_{31}H_{33}O_{13}$ were dereplicated as actinosporin A and C, respectively, which were discovered from the culture of *Actinokineospora spheciospongiae* strain EG49 [23,24]. The formula $C_{21}H_{18}O_8$ was dereplicated as daunomycinone, which was reported from *Streptomyces coeruleorubid* [25].

The formulas $C_{26}H_{25}O_{11}$ and $C_{25}H_{24}O_8$ were dereplicated as atramycin A and B, respectively. These isotetracenone metabolites were discovered from *Streptomyces atratus* BY90 [26]. Additionally, the suggested molecular formula $C_{18}H_{12}O_5$ was dereplicated as lagumycin B, which was previously isolated from *Micromonospora* sp. [27], while the formula $C_{16}H_{12}O_5$ was dereplicated as the isoflavonoid kakkatin that was reported from the soil-derived *Streptomyces* strain YIM GS3536. Moreover, it was discovered in another terrestrial *Streptomyces* sp. GW39/1530 [28,29]. Furthermore, the molecular formula $C_9H_9NO_3$ was dereplicated as erbstatin, a simple dehydrotyrosine derivative isolated from *Streptomyces amnkusaensis* [30,31]. Additionally, the molecular formula $C_{36}H_{48}N_2O_8$ was dereplicated as ansatrienin A, previously detected in *Streptomyces collinus* [32]. Moreover, the formulas $C_{25}H_{47}N_5O_4$, $C_{26}H_{49}N_5O_4$, and $C_{28}H_{53}N_5O_4$ were dereplicated as cyclic tetrapeptides rhodopeptin C1, C2, and B5, respectively, which were formerly reported in *Rhodococcus* sp. [33,34]. The formula $C_{32}H_{48}N_6O_9$ was dereplicated as the peptide actinoramide B, which was detected in a marine bacterium highly corelated to the genus *Streptomyces* [35]. Likewise, the formula $C_{17}H_{26}O_4$ was dereplicated as cineromycin-B antibiotic that showed significant MRSA inhibition, which was isolated from the actinomycetales strain INA 2770 [36]. The formula $C_{19}H_{27}N_5O_7$ was annotated as heterobactin B, a siderophore discovered from *Rhodococcus erythropolis* IGTS8 [37], while the formula $C_{26}H_{39}NO_5$ was dereplicated as piericidin-F, which was reported from *Streptomyces* sp. CHQ-64 [38]. Additionally, the formula $C_{27}H_{39}NO_7$ was annotated as migrastatin, which was reported as a tumour cell migration inhibitor and isolated from *Streptomyces* sp. MK929-43F1 [39]. Moreover, the formula $C_{24}H_{46}N_6O_8$ was dereplicated as proferrioxamine-A1, a siderophore isolated from *Streptomyces xinghaiensis* NRRL B-24674T [40]. Furthermore, the formula $C_{23}H_{38}O_5$ was dereplicated as the 16-membered lactone protylonolide, which was identified as the metabolite of mycaminose idiotroph that has been obtained from *Streptomyces fradiae* KA-427 [41]. Moreover, the formula $C_{37}H_{62}O_{11}$ was dereplicated as the polyether 26-deoxylaidlomycin isolated from *Streptoverticillium olivoreticuli* IMET 43,782 [42], while the suggested formula $C_{35}H_{58}O_{10}$ was dereplicated as macrolide kaimonolide B, which was discovered in *Streptomyces* sp. no. 4155 and shown to significantly inhibit plant growth [43]. Furthermore, the formula $C_{25}H_{44}O_7$ was dereplicated as 8,15-dideoxylankanolide, which was reported in *Streptomyces rochei* 7434AN4 [44]. The molecular formula $C_{34}H_{60}O_{10}$ was identified as the polyether antibiotic ferensimycin-A, previously discovered in *Streptomyces* sp. no. 5057 [45]. Likewise, the formula $C_{26}H_{46}N_6O_5$ was identified as the cytotoxic peptide lucentamycin C, which was reported from a marine-derived actinomycete *Nocardiopsis lucentensis* CNR-712 [46]. Finally, the formula $C_{50}H_{92}O_{14}$ was dereplicated as glucolipsin-A, a glucokinase activator that has been isolated from *Streptomyces puvpuvogenisclevoticus* [47].

It is worth noting that the compounds listed in Table 1 were traced in the LC–HRESIMS analysis of the coculture extract. The producing strain for each compound was predicted on the basis of literature. However, mitomycin-K, 8,15-dideoxylankanolide, piericidin-F, migrastatin, kaimonolide B, rhodopeptin C1, rhodopeptin C2, and rhodopeptin B5 were also traced in the axenic culture of *Rhodococcus* sp. UR59. Additionally, actinosporins A and C, and UK-2B were also traced in the axenic culture of *Actinokineospora spheciospongiae* strain EG49. All other reported metabolites in Table 1 were not traced in the axenic cultures and were induced during the coculture fermentation.

Table 1. Metabolomics analysis of the coculture extract of *Actinokineospora spheciospongiae* strain EG49 and *Rhodococcus* sp. UR59.

Rt (min)	*m/z* [M − H]−	*m/z* [M + H]+	Molecular Formula	Tentative Identification	Strain EG49	Strain UR59	Coculture	Bioactivity	Ref.
2.47		303.1341	$C_{16}H_{18}N_2O_4$	Mitomycin-K	-	+	+	antitumor	[20]
2.91		327.0866	$C_{18}H_{14}O_6$	Fluostatin-B	-	-	+	antinociceptive	[22]
2.94	657.1821		$C_{32}H_{33}O_{15}$	Actinosporin A	+	-	+	anti-trypanosomal	[23]
2.96		613.1926	$C_{31}H_{33}O_{13}$	Actinosporin C	+	-	+	antioxidant	[24]
2.99		399.1075	$C_{21}H_{18}O_8$	Daunomycinone	-	-	+	-	[48]
3.04		599.2125	$C_{31}H_{34}O_{12}$	Actinosporin F	-	-	+	-	[49]
3.08		469.1492	$C_{25}H_{24}O_9$	Actinosporin E	-	-	+	-	[49]
3.11		467.1336	$C_{25}H_{22}O_9$	Actinosporin H	-	-	+	-	[49]
3.15	513.1399		$C_{26}H_{25}O_{11}$	Atramycin A	-	-	+	antitumor	[26]
3.25		451.1389	$C_{25}H_{22}O_8$	Actinosporin G	-	-	+	-	[49]
3.30		309.0757	$C_{18}H_{12}O_5$	Lagumycin B	-	-	+	anticancer	[27]
3.76	178.0499		$C_9H_9NO_3$	Erbstatin	-	-	+	anticancer	[30]
3.81	635.3315		$C_{36}H_{48}N_2O_8$	Ansatrienin A	-	-	+	antifungal	[32]
3.93	192.0655		$C_{10}H_{11}NO_3$	Spoxazomicin C	-	-	+	anti-trypanosomal	[50]
4.12		661.3568	$C_{32}H_{48}N_6O_9$	Actinoramide B	-	-	+	antimalarial	[51]
4.17	293.1749		$C_{17}H_{26}O_4$	Cineromycin-B	-	-	+	antibacterial	[52]
4.54		438.1974	$C_{19}H_{27}N_5O_7$	Heterobactin B	-	-	+	siderophore	[37]
6.41	451.1391		$C_{25}H_{24}O_8$	Atramycin B	-	-	+	antitumor	[26]
6.34	444.2744		$C_{26}H_{39}NO_5$	Piericidin-F	-	+	+	anticancer	[38]
6.91	488.2649		$C_{27}H_{39}NO_7$	Migrastatin	-	+	+	anticancer	[53]
7.24		547.3455	$C_{24}H_{46}N_6O_8$	Proferrioxamine-A1	-	-	+	siderophore	[40]
7.40	393.2640		$C_{23}H_{38}O_5$	Protylonolide	-	-	+	antibiotic	[41]
7.58		683.4347	$C_{37}H_{62}O_{11}$	26-Deoxylaidlomycin	-	-	+	antibacterial	[54]
7.76		482.3687	$C_{25}H_{47}N_5O_4$	Rhodopeptin C1	-	+	+	Antifungal	[33]
8.15		639.4084	$C_{35}H_{58}O_{10}$	Kaimonolide B	-	+	+	plant growth inhibitor	[43]
9.09		496.3846	$C_{26}H_{49}N_5O_4$	Rhodopeptin C2	-	+	+	antifungal	[34]
9.14		524.4156	$C_{28}H_{53}N_5O_4$	Rhodopeptin B5	-	+	+	antifungal	[34]
9.80		315.0865	$C_{17}H_{14}O_6$	Capillasterquinone B	-	-	+	NO production inhibitor	[55]
9.81		305.0810	$C_{19}H_{12}O_4$	Tetrangulol		-	+	antibiotic	[56]
9.98		457.3141	$C_{25}H_{44}O_7$	8,15-Dideoxylankanolide	-	+	+	-	[44]
10.22		527.2022	$C_{27}H_{31}N_2O_9$	UK-2B	+	-	+	antifungal	[57]
11.01		629.4242	$C_{34}H_{60}O_{10}$	Ferensimycin-A	-	-	+	antibiotic	[45]
11.23		523.3601	$C_{26}H_{46}N_6O_5$	Lucentamycin C	-	-	+	anticancer	[46]
11.28		917.6546	$C_{50}H_{92}O_{14}$	Glucolipsin-A	-	-	+	glucokinase activator	[47]

2.3. Identification of the Isolated Compounds (**1**–**8**)

Chemical structures of the purified metabolites **1–8** from the coculture were assigned on the basis of comparing the LC–HRESIMS analysis, 1D and 2D NMR spectral data, and optical rotation measurements to the published literature (Figure 2). Accordingly, compounds **1–3** have been previously isolated from *Actinokineospora spheciospongiae* strain EG49 and identified as the angucyclinone antibiotics actinosporin E, H, and G, respectively, through the activation of their cryptic gene cluster by N-acetylglucosamine [50]. Compound **4** was assigned as spoxazomicin C of the pyochelin family of antibiotics, which was previously isolated from the culture broth of the endophyte *Streptosporangium oxazolinicum* K07-0460T [51]. In contrast, compound **5** was previously identified as the angucyclinone antibiotic tetrangulol, which was previously isolated from *Streptomyces rimosus* [58] and recently from *Amycolatopsis* sp. HCa1 [59]. Compound **6** was previously discovered as capillasterquinone B, an anthraquinone that was isolated from the crinoid *Capillaster multiradiatus* [57]. Moreover, compound **7** was identified as L-tryptophanamide. We propose it as an artefact as it was not traced in the LC–MS analysis of either the axenic or the coculture extracts, and thus it was probably generated during the fractionation and purification process. Finally, compound **8** was isolated from *Streptomyces* sp. 517-02 [57] and identified as UK-2B, an antifungal antibiotic with similarity in structure to antimycin A [60].

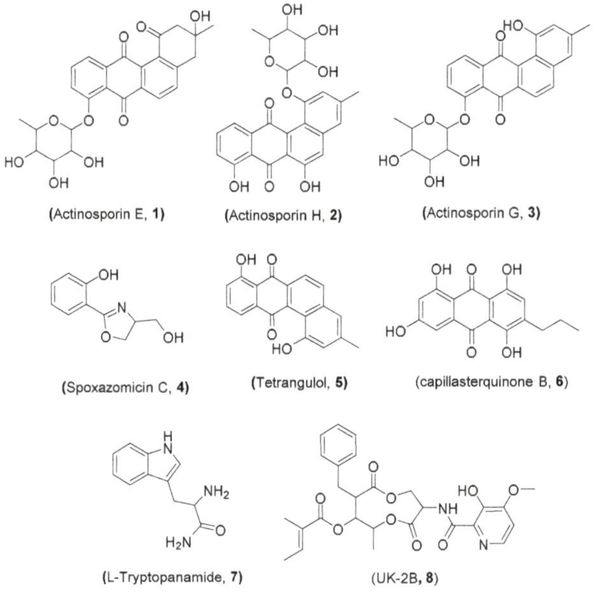

Figure 2. Compounds isolated from the coculture of *Actinokineospora spheciospongiae* strain EG49 and *Rhodococcus* sp. UR59.

The same *Actinokineospora spheciospongiae* strain EG49 was subjected to N-acetyl-D-glucosamine (GluNAc)-mediated silent gene activation to produce new actinosporins E–H, the same actinosporins E (**1**), G (**3**), and H (**2**) discussed here under coculture and aglycone angucycline tetrangulol (**5**), which was not reported from the axenic culture treated with GluNAc [50]. The amino sugar GluNAc is a signalling molecule that can induce microbial secondary metabolism. It is present as a cell wall component in peptidoglycan or chitin in the bacterial or fungal cell wall, respectively [60,61]. Having observed a similar induction when *Actinokineospora spheciospongiae* strain EG49 was cocultured with *Rhodococcus* sp. UR59, we can assume that *Rhodococcus* sp. UR59 directed the biosynthesis of actinosporins in a similar way to GluNAc. This could be as exudation of GluNAc by one of the species into the coculture environment to trigger antibiotic production more likely from *Rhodococcus* sp. UR59 as defence molecules. The studies further support this and demonstrated that GluNAc is secreted by bacteria under malnourished conditions to signal antibiotic production against opposite competitors in the vicinity [62]. However, this requires further studies on the coculture medium to identify excreted GluNAc or compounds with similar signalling function.

2.4. Antimalarial Screening

Angucyclines are microbial secondary metabolites known as promising antimicrobial, anticancer, and antimalarial agents [63–65]. The core structure of angucyclines is characterised by a benz[α]anthracene ring, an angular tetracycline ring system [60]. The reported angucyclines can be categorised as aglycones such as saccharosporones A, B, and C [60], and glycosylated angucyclines such as pseudonocardones A−C [63] and urdamycinone E, urdamycinone G, and dehydroxyaquayamycin isolated from fungal and bacterial strains [62]. However, different antimalarial activity profiles between aglycones and glycosylated angucyclines have not been explained.

The potential antiparasitic effectiveness of the angucycline scaffold and the promising antimalarial effect exhibited by the total extract of the coculture of *Actinokineospora spheciospongiae* strain EG49 and *Rhodococcus* sp. UR59 (IC$_{50}$ value of 0.13 µg/mL, Table 2) when screened against *Plasmodium falciparum* have encouraged us to perform large-scale coculture fermentation. Large-scale fermentation followed by liquid–liquid fractionation and HPLC purification of the active sub-fraction led to the isolation of eight metabolites. The antimalarial screening of the isolated compounds indicated that the angucycline glycosides **1–3** and aglycone **5** and the anthraquinone **6** exhibited antimalarial effect with IC$_{50}$ values in the range of 9–13.5 µg/mL in comparison to the IC$_{50}$ value of the positive control chloroquine (0.022 µg/mL). The activity of the compounds **1–3**, **5**, and **6** was further studied by docking against a few known drug targets to suggest these compounds as potential leads to be developed for enhanced activity. It worth noting that the isolated molecules did not show the expected antimalarial activity, which could be attributed to either the synergistic effect of microbial metabolites in the coculture extract or the presence of minor molecules that were too scarce to be isolated even after large-scale fermentation.

Table 2. Antimalarial effect of the bacterial coculture derived metabolites.

Compound	IC$_{50}$ Values (µg/mL) [1]
Coculture extract	0.13
1	12.6
2	13.6
3	11.2
4	>50
5	9.7
6	9.2
7	>50
8	>50
Chloroquine	0.022

[1] Average of two independent runs.

2.5. Docking Analysis

Compounds (**1–3**, **5**, **6**) that showed inhibitory activity against *P. falciparum* were subjected to molecular docking experiments against a number of reported malaria targets, e.g., NADH:ubiquinone oxidoreductase (PDB: 5JWA), Kelch protein (PDB: 4YY8), *P. falciparum* protein kinase (PDB: 1V0P), NADH dehydrogenase 2 (PDB:4PD4), and lysyl-tRNA synthetase (PDB:6AGT). They achieved the best scores (binding energy −8.5 to −9.1 kcal/mol) against the later target, lysyl-tRNA synthetase (PfKRS1). Moreover, they exhibited binding mode inside the active site compared to the co-crystalised ligand [66]. As shown in Table 3 and Figure 3, these compounds exhibited multiple interactions with several amino acids inside the enzyme's active site, where ARG-330, HIS-338, GLU-500, ARG-559, and PHE-342 were the most common interacting ones. Hence, this attractive scaffold can be utilised in the future design of antimalarial therapeutics targeting PfKRS1 (Table 3). Antimalarial effect of the bacterial coculture derived metabolites.

Table 3. Binding scores and interacting amino acid residues with compounds **1–3**, **5**, and **6** inside the lysyl-tRNA synthetase (PfKRS1)'s active site.

Compound	Binding Energy (kcal/mol)	H-Bonding	Hydrophobic Interactions
1	−8.9	ARG-330	ALA-446, GLU-500, LYS-607
2	−8.3	ASP-450, SER-454, MET-475	ARG-330, HIS-338, ASP-450, GLU-458, GLU-500
3	−10.3	GLU-493	ARG-330, PHE-342, ASP-450, ARG-559
5	−9.1	GLU-500, THR-337	ARG-330, HIS-338, PHE-342, ARG-559
6	−9.0	ARG-330, ASN-339	ARG-330, HIS-338, PHE-342, GLU-500, ARG-559, LYS-607
Co-crystalised ligand	−9.5	ASN-339, GLY-556	ARG-330, HIS-338, PHE-342

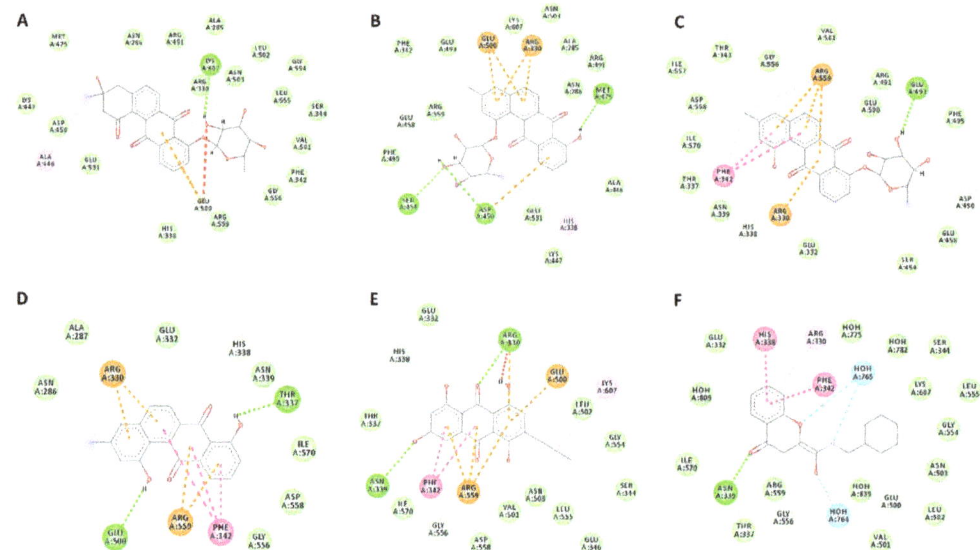

Figure 3. Binding modes of compounds **1–3**, **5**, and **6** ((**A**–**E**) respectively) together with the co-crystalised ligand (**F**) inside PfKRS's active site. Dashed lines indicate interactions between each ligand and the active site's amino acid residues. Green colour indicates H-bonding; orange colour indicates π–anion or π–cation interactions; pink colour indicates hydrophobic interactions.

3. Materials and Methods

3.1. General Experimental

Extract purification was conducted by preparative Agilent 1100 series HPLC equipped with gradient pump and DAD using a reversed-phase Sunfire (C18, 5 μm, 10 × 250 mm, serial no. 226130200125). All 1D and 2D NMR spectral data were acquired using a JEOL ECZ-R500 NMR spectrometer equipped with a Royal 5 mm combined broadband and inverse probe. Thermo LTQ Orbitrap coupled to an HPLC system was utilised to acquire HRESIMS data using capillary temperature of 260 °C, capillary voltage of 45 V, sheath gas flow rate of 40–50 arbitrary units, auxiliary gas flow rate of 10–20 arbitrary units, spray voltage of 4.5 kV, and mass range of 100–2000 amu (maximal resolution of 60,000). Optical rotations and UV spectra acquisition were acquired using a Perkin-Elmer 343 polarimeter and Perkin-Elmer Lambda2 UV–VIS spectrometer, respectively.

3.2. Actinomycetes Isolation

Callyspongia sp. was collected from Hurghada (Red Sea, Egypt) at a depth of 5 m and latitude 27°17′01.0″ N and longitude 33°46′21.0″ E. The sponge specimen was identified by Prof. El-Sayd Abed El-Aziz (Department of Invertebrates Lab., National Institute of Oceanography and Fisheries, Egypt). The sponge was transported in a plastic bag in seawater to the laboratory and washed thoroughly with sterile seawater. The surface sterilised specimen was cut into pieces of ≈1 cm^3, followed by vigorous homogenising with 10 volumes of sterile seawater in a pre-sterilised mortar. Serially diluted supernatant (10^{-1}, 10^{-2}, 10^{-3}) was subsequently plated on to the sterile agar plates. For the isolation of different actinomycetes, we used M1, ISP2, and marine agar (MA) media were used [18]. The isolation of slow-growing actinomycetes was performed by supplementing all media with filtered 25 μg/mL nalidixic acid, 25 μg/mL nystatin, and 100 μg/mL cycloheximide. The inoculated plates were stored in an incubator for 6–8 weeks at 30 °C. Subculturing of distinct colony morphotypes resulted in pure strains. *Rhodococcus* sp. UR59 was cultured on ISP2 medium and preserved in 20% glycerol at −80 °C. On the other hand,

Actinokineospora spheciospongiae strain EG49 was previously recovered and identified from the Red Sea sponge *Spheciospongia vagabunda* [18].

3.3. Molecular Identification and Phylogenetic Analysis

With reference to Hentschel et al., we carried out 16S rRNA gene amplification, cloning, and sequencing using 27F and 1492RRNA as universal primers [18]. By using the Pintail programme, we identified chimeric sequences [67]. The sequence's genus level affiliation was validated using the Project Classifier of the Ribosomal Database. All the sequences were classified at the genus level by the RDP Classifier (g 16srrna, f allran) and confirmed with the SILVA Incremental Aligner (SINA) [68]. Using the SINA Web Aligner, an alignment was determined again (variability profile: bacteria). The Gap-only position with trimALL was eliminated (-noallgaps). The best fitting model was initially calculated for phylogenetic tree construction with the Model Generator. To produce the phylogenetic tree, we applied RAxML (-f a-m GTRGAMMA-x 12345-p 12345 -# 1000) and the estimated model with 1000 bootstrap resamples. With Interactive Tree of Life (ITOL) [69], visualisation was achieved. The BLAST with the accession number MW453143 was deposited at Genebank.

3.4. Co-Cultivation and Extract Preparation

Rhodococcus sp. UR59 and *Actinokineospora spheciospongiae* strain EG49 were cultivated on liquid media M1 and ISP2 as axenic and cocultures. A total of 20 mL of 3-day-old culture of *Rhodococcus* sp. was used for large scale fermentation. *Rhodococcus* sp. UR59 was transferred to 20 × 2 L Erlenmeyer flasks containing 1 L of ISP2 medium pre-inoculated with 20 mL of 4-day-old *Actinokineospora spheciospongiae* strain EG49 and left for 7 days at 25 °C and 180 rpm in a shaker incubator. After fermentation, the culture was filtered, and the supernatant was extracted twice with ethyl acetate (1.5 L each) followed by evaporation under vacuum to provide the ethyl acetate extract (850 mg).

3.5. Metabolic Profiling

For mass spectrometry analysis, the dry ethyl acetate extracts from different microbial and coculture samples were dissolved in MeOH at 1 mg/mL and subjected to metabolic analysis using LC–HRESIMS according to Abdelmohsen et al. [23]. An Acquity UPLC system coupled to a Synapt G2 HDMS qTOF hybrid mass spectrometer (Waters, Milford, CT, USA) was used to acquire the HRMS data using capillary temperature at 320 °C, spray voltage at 4.5 kV, and mass range of *m/z* 150–1500; both positive and negative ESI modes were applied. The MS was processed using MZmine 2.20 on the basis of the defined parameters [23]. The chromatogram builder and chromatogram deconvolution were detected and followed by mass ion peaks. The isotopes were differentiated by grouper isotopic peaks and the missing peaks were depicted using the gap-filling peak finder. Then, molecular formula prediction and peak identification were conducted from the processed positive and negative ionisation mode datasets. Finally, the peaks were dereplicated against the Dictionary of Natural Products (DNP) database.

3.6. Metabolites Isolation

The crude co-fermentation ethyl acetate (EtOAc) (850 mg) was chromatographed on Sephadex LH-20 (32–64 μm, 100 × 25 mm) column using an 80:20 MeOH/H_2O eluent in order to obtain 5 fractions (Fr.1–Fr.6). The third bioactive fraction (300 mg) was then chromatographed using silica gel column with a gradient elution starting at DCM/EtOAc (100:0 to 0:100) then 100% MeOH to obtain 8 sub-fractions. The active subfractions 4 and 5 were combined (85 mg) and further subjected to semi preparative HPLC purification (Sunfire, C18, 5 μm, 10 × 250 mm) with a gradient of 20%–100% CH_3CN in H_2O over 30 min and 10 min at 100% CH_3CN at 1.5 mL/min flow rate to yield compound **7** (t_R 9.6 min, 7.5 mg), **2** (t_R 10.7 min, 4.5 mg), **3** (t_R 11.2 min, 2.5 mg), **4** (t_R 15.2 min, 2.8 mg), **5** (t_R 18.3 min, 2.1 mg), **1** (t_R 24.6 min, 3.2 mg), **6** (t_R 27.2 min, 3.8 mg), and **8** (t_R 31.3 min, 1.5 mg).

3.7. Antimalarial Screening

The Malstat assay was used as mentioned earlier to assess the compounds' antimalarial effect [70,71]. The compounds were dissolved in DMSO (Sigma Aldrich, Taufkirchen, Germany) at concentrations ranging from 50 µg/mL to 0.4 µg/mL, and synchronised *P. falciparum* 3D7 ring stage cultures were placed in duplicate at a parasite level of 1% in 96-well plates (200 µL/well). Chloroquine (CQ; Sigma Aldrich, Taufkirchen, Germany) was used as a positive control. The *P. falciparum* 3D7 parasite was cultured with the compounds at 37 °C in 5% O_2, 5% CO_2, and 90% N_2 for 72 h. After this, 20 µL was transferred to 100 µL of the Malstat reagent (0.1% Triton X-100, 1 g of L-lactate, 0.33 g Tris, and 33 mg of APAD (3-acetylpyridine adenine dinucleotide; Taufkirchen, Germany)) dissolved in 100 mL of distilled water (pH 9.0) in a 96-well microtiter plate. The plasmodial lactate dehydrogenase (LDH) activity was then evaluated by adding to the Malstat reaction 20 µL of a 1:1 mixture of diaphorase (1 mg/mL) and nitro blue tetrazolium (NBT). The optical densities were estimated at 630 nM, and the IC_{50} values were determined using the GraphPad Prism software version 5 from variable-slope sigmoidal dose–response curves (GraphPad Software Inc., La Jolla, CA, USA).

3.8. Molecular Docking

Docking analysis was carried out using the Discovery Studio 2.5 software (Accelrys Inc., San Diego, CA, USA). Completely automatic docking tool using "Dock ligands (CDOCKER)" procedure operating on Intel Core i32370 CPU @ 2.4 GHz 2.4 GHz, RAM Memory 2 GB under the Windows 10.0 system. Furthermore, these docked compounds were assembled using a software Chem 3D ultra 12.0 (Cambridge Soft Corporation, USA (2010)), and then sent to the Discovery Studio 2.5 software. From this, an automatic protein formulation procedure was conducted through the MMFF94 forcefield with the binding site sphere recognised by the software. The receptor was recorded as "input receptor molecule" in the CDOCKER protocol explorer. Establishing this, the test compounds were subjected to force fields to obtain the minimum energy structure. These poses were ranked and studied thoroughly, showing the best ligand–HDAC interactions from the calculations and 2D and 3D examinations [72,73].

4. Conclusions

Microbial coculture continues to prove its efficiency in triggering the production of cryptic microbial secondary metabolites. Mixed cultivation of two Red Sea-derived actinobacteria, namely, *Actinokineospora spheciospongiae* strain EG49 and *Rhodococcus* sp. UR59, resulted in the induction of several non-traced metabolites in their axenic cultures. Interestingly, actinosporins E–H were reported to be induced when the axenic culture of the *Actinokineospora spheciospongiae* strain EG49 was treated with the signalling molecule GluNAc. Such induction was comparable to that made by the *Rhodococcus* sp. UR59 in the coculture environment, providing the effectiveness of co-cultivation in the discovery of microbial metabolites yet to be discovered in the axenic fermentation with the potential that could be comparable to adding signalling molecules in the fermentation flask. Additionally, the induced actinosporins exhibited a promising antimalarial effect that is likely to be through the inhibition of *P. falciparum* lysyl-tRNA synthetase, which requires further investigation as an interesting structural motif for the development of new antimalarial therapeutics.

Author Contributions: Conceptualisation, H.A.A., B.T., U.R.A., and M.E.R.; methodology, H.A.A., B.T., M.H.A.H., F.A.B., C.J.N., and H.M.H.; software, H.A.A. and B.T.; validation, G.P., U.R.A., and M.E.R.; formal analysis, H.A.A., B.T., M.H.A.H., F.A.B., C.J.N., and H.M.H.; data curation, H.A.A. and B.T.; writing—original draft preparation, H.A.A., B.T., M.H.A.H., F.A.B., C.J.N., and H.M.H.; writing—review and editing, G.P., U.R.A., and M.E.R.; supervision, G.P., U.R.A., and M.E.R.; project administration, M.E.R.; funding acquisition, H.A.A. All authors have read and agreed to the published version of the manuscript.

Funding: This research was funded by the Deanship of Scientific Research (DSR), King Abdulaziz University, Jeddah, under grant no. (DF-316-142-1441). The authors, therefore, gratefully acknowledge DSR technical and financial support.

Institutional Review Board Statement: Not applicable.

Data Availability Statement: No supplementary data available with this article.

Acknowledgments: The authors also like to thank the Deanship of Scientific Research (DSR), King Abdulaziz University, for their technical and financial support.

Conflicts of Interest: The authors declare no conflict of interest.

References

1. Nai, C.; Meyer, V. From axenic to mixed cultures: Technological advances accelerating a paradigm shift in microbiology. *Trends Microbiol.* **2018**, *26*, 538–554. [CrossRef]
2. Wu, C.; Choi, Y.H.; van Wezel, G.P. Metabolic profiling as a tool for prioritising antimicrobial compounds. *J. Ind. Microbiol.* **2006**, *43*, 299–312.
3. Watve, M.G.; Tickoo, R.; Jog, M.M.; Bhole, B.D. How many antibiotics are produced by the genus Streptomyces? *Arch. Microbiol.* **2001**, *176*, 386–390. [CrossRef] [PubMed]
4. Harvey, A.L.; Edrada-Ebel, R.; Quinn, R.J. The re-emergence of natural products for drug discovery in the genomics era. *Nat. Rev. Drug Discov.* **2015**, *14*, 111–129. [CrossRef]
5. Sanchez, J.F.; Somoza, A.D.; Keller, N.P.; Wang, C.C. Advances in Aspergillus secondary metabolite research in the post-genomic era. *Nat. Prod. Rep.* **2012**, *29*, 351–371. [CrossRef] [PubMed]
6. Inglis, D.O.; Binkley, J.; Skrzypek, M.S.; Arnaud, M.B.; Cerqueira, G.C.; Shah, P.; Wymore, F.; Wortman, J.R.; Sherlock, G. Comprehensive annotation of secondary metabolite biosynthetic genes and gene clusters of Aspergillus nidulans, A. fumigatus, A. niger and A. oryzae. *BMC Microbiol.* **2013**, *13*, 91. [CrossRef]
7. Brakhage, A.A. Regulation of fungal secondary metabolism. *Nat. Rev. Microbiol.* **2013**, *11*, 21–32. [CrossRef]
8. Bednarski, Z.; Bednarska, H. First research work by Robert Koch on etiology of anthrax-in cooperation with Józef Knechtel, Polish apothecary. *Arch. Hist. Filoz. Med.* **2003**, *66*, 161–168.
9. Fischer, A. Acid production graphically registered as an indicator of the vital processes in the cultivation of bacteria. *J. Exp. Med.* **1918**, *28*, 529–545. [CrossRef]
10. Degenkolb, T.; Heinze, S.; Schlegel, B.; Strobel, G.; Gräfe, U. Formation of new lipoaminopeptides, acremostatins A, B, and C, by co-cultivation of *Acremonium* sp. Tbp-5 and Mycogone rosea DSM 12973. *Biosci. Biotechnol. Biochem.* **2002**, *6*, 883–886. [CrossRef] [PubMed]
11. Wakefield, J.; Hassan, H.M.; Jaspars, M.; Ebel, R.; Rateb, M.E. Dual induction of new microbial secondary metabolites by fungal bacterial co-cultivation. *Front. Microbiol.* **2017**, *8*, 1284. [CrossRef] [PubMed]
12. Li, C.; Sarotti, A.M.; Yang, B.; Turkson, J.; Cao, S. A new N-methoxypyridone from the co-cultivation of Hawaiian endophytic fungi Camporesia sambuci FT1061 and Epicoccum sorghinum FT1062. *Molecules* **2017**, *22*, 1166. [CrossRef]
13. Thissera, B.; Alhadrami, H.A.; Hassan, M.H.; Hassan, H.M.; Bawazeer, M.; Yaseen, M.; Belbahri, L.; Rateb, M.E.; Behery, F.A. Induction of cryptic antifungal pulicatin derivatives from Pantoea agglomerans by microbial coculture. *Biomolecules* **2020**, *10*, 268. [CrossRef] [PubMed]
14. Dobson, M.J. Exploring natural remedies from the past. *Parassitologia* **1998**, *40*, 69–81.
15. Pyae Phyo, A.; Nkhoma, S.; Singhasivanon, P.; Day, N.P.; White, N.J. Emergence of artemisinin-resistant malaria on the western border of Thailand: A longitudinal study. *Lancet* **2012**, *379*, 1960–1966. [CrossRef]
16. Straimer, J.; Gnädig, N.F.; Witkowski, B.; Amaratunga, C.; Duru, V.; Ramadani, A.P.; Dacheux, M.; Khim, N.; Zhang, L.; Lam, S.; et al. K13-propeller mutations confer artemisinin resistance in Plasmodium falciparum clinical isolates. *Science* **2015**, *347*, 428–431. [CrossRef] [PubMed]
17. Yeka, A.; Lameyre, V.; Afizi, K.; Fredrick, M.; Lukwago, R.; Kamya, M.R.; Talisuna, A.O. Efficacy and safety of fixed-dose artesunate-amodiaquine vs. artemether-lumefantrine for repeated treatment of uncomplicated malaria in Ugandan children. *PLoS ONE* **2014**, *9*, e113311. [CrossRef]
18. Abdelmohsen, U.R.; Pimentel-Elardo, S.M.; Hanora, A.; Radwan, M.; Abou-El-Ela, S.H.; Ahmed, S.; Hentschel, U. Isolation, phylogenetic analysis and anti-infective activity screening of marine sponge-associated actinomycetes. *Mar. Drugs* **2010**, *8*, 399–412. [CrossRef] [PubMed]
19. Kampfer, P.; Glaeser, S.P.; Busse, H.J.; Abdelmohsen, U.R.; Ahmed, S.; Hentschel, U. *Actinokineospora spheciospongiae* sp. nov., isolated from the marine sponge *Spheciospongia vagabunda*. *Int. J. Syst. Evol. Microbiol.* **2015**, *65 Pt 3*, 879–884. [CrossRef]
20. Bass, P.D.; Gubler, D.A.; Judd, T.C.; Williams, R.M. Mitomycinoid alkaloids: Mechanism of action, biosynthesis, total syntheses, and synthetic approaches. *Chem. Rev.* **2013**, *113*, 6816–6863. [CrossRef]

21. Gu, Q.S.; Yang, D. Enantioselective Synthesis of (+)-Mitomycin K by a Palladium-Catalysed Oxidative Tandem Cyclization. *Angew. Chem.* **2017**, *129*, 5980–5983. [CrossRef]

22. Akiyama, T.; Harada, S.; Kojima, F.; Takahashi, Y.; Imada, C.; Okami, Y.; Muraoka, Y.; Aoyagi, T.; Takeuchi, T. Fluostatins A and B, New Inhibitors of Dipeptidyl Peptidase III, Produced by Streptomyces sp. TA-3391. *J. Antibiot.* **1998**, *51*, 553–559. [CrossRef]

23. Abdelmohsen, U.R.; Cheng, C.; Viegelmann, C.; Zhang, T.; Grkovic, T.; Ahmed, S.; Quinn, R.J.; Hentschel, U.; Edrada-Ebel, R. Dereplication strategies for targeted isolation of new antitrypanosomal actinosporins A and B from a marine sponge associated-*Actinokineospora* sp. EG49. *Mar. Drugs* **2014**, *12*, 1220–1244. [CrossRef]

24. Grkovic, T.; Abdelmohsen, U.R.; Othman, E.M.; Stopper, H.; Edrada-Ebel, R.; Hentschel, U.; Quinn, R.J. Two new antioxidant actinosporin analogues from the calcium alginate beads culture of sponge-associated *Actinokineospora* sp. strain EG49. *Bioorg. Med. Chem. Lett.* **2014**, *24*, 5089–5092. [CrossRef] [PubMed]

25. Blumauerova, M.; Kralovcova, E.; Matějů, J.; Jizba, J.; Vaněk, Z. Biotransformations of anthracyclinones in *Streptomyces coeruleorubidus* and *Streptomyces galilaeus*. *Folia Microbiol.* **1979**, *24*, 117. [CrossRef]

26. Fujioka, K.; Furihata, K.; Shimazu, A.; Hayakawa, Y.; Seto, H. Isolation and characterisation of atramycin A and atramycin B, new isotetracenone type antitumor antibiotics. *J. Antibiot.* **1991**, *44*, 1025–1028. [CrossRef]

27. Mullowney, M.W.; Ó hAinmhire, E.; Tanouye, U.; Burdette, J.E.; Pham, V.C.; Murphy, B.T. A Pimarane Diterpene and Cytotoxic Angucyclines from a Marine-Derived Micromonospora sp. in Vietnam's East Sea. *Mar. Drugs* **2015**, *13*, 5815–5827. [CrossRef] [PubMed]

28. Huang, R.; Ding, Z.G.; Long, Y.F.; Zhao, J.Y.; Li, M.G.; Cui, X.L.; Wen, M.L. A new isoflavone derivative from *Streptomyces* sp. YIM GS3536. *Chem. Nat. Compd.* **2013**, *48*, 966–969. [CrossRef]

29. Maskey, R.P.; Asolkar, R.N.; Speitling, M.; Hoffman, V.; Grün-Wollny, I.; Fleck, W.F.; Laatsch, H. Flavones and new isoflavone derivatives from microorganisms: Isolation and structure elucidation. *Z. Naturforsch. B* **2003**, *58*, 686–691. [CrossRef]

30. Sugumaran, M.; Robinson, W.E. Bioactive dehydrotyrosyl and dehydrodopyl compounds of marine origin. *Mar. Drugs* **2010**, *8*, 2906–2935. [CrossRef]

31. Umezawa, H.; Imoto, M.; Sawa, T.; Isshiki, K.; Matsuda, N.; Uchida, T.; Iinuma, H.; Hamada, M.; Takeuchi, T. Studies on a new epidermal growth factor-receptor kinase inhibitor, erbstatin, produced by MH435-hF3. *J. Antibiot.* **1986**, *39*, 170–173. [CrossRef]

32. Weber, W.; Zähner, H.; Damberg, M.; Russ, P.; Zeeck, A. Metabolic products of microorganisms 201. Ansatrienin A and B, antifungal antibiotics from Streptomyces collinus. Zentralblatt für Bakteriologie Mikrobiologie und Hygiene: I. Abt. *Orig. C Allg. Angew. Okol. Mikrobiol.* **1981**, *2*, 122–139.

33. Chiba, H.; Agematu, H.; Kaneto, R.; Terasawa, T.; Sakai, K.; Dobashi, K.; Yoshioka, T. Rhodopeptins (Mer-N1033), Novel Cyclic Tetrapeptides with Antifungal Activity from *Rhodococcus* sp. *J. Antibiot.* **1999**, *52*, 695–699. [CrossRef]

34. Nakayama, K.; Kawato, H.C.; Inagaki, H.; Nakajima, R.; Kitamura, A.; Someya, K.; Ohta, T. Synthesis and antifungal activity of rhodopeptin analogues. 2. Modification of the west amino acid moiety. *Org. Lett.* **2000**, *2*, 977–980. [CrossRef]

35. Nam, S.J.; Kauffman, C.A.; Jensen, P.R.; Fenical, W. Isolation and characterization of actinoramides A–C, highly modified peptides from a marine *Streptomyces* sp. *Tetrahedron* **2011**, *67*, 6707–6712. [CrossRef] [PubMed]

36. Terekhova, L.P.; Galatenko, O.A.; Kulyaeva, V.V.; Malkina, N.D.; Boikova, Y.V.; Katrukha, G.S.; Shashkov, A.S.; Gerbst, A.G.; Nifantiev, N.E. Isolation, NMR spectroscopy, and conformational analysis of the antibiotic ina 2770 (cineromycin B) produced by Streptomyces strain. *Russ. Chem.* **2007**, *56*, 815–818. [CrossRef]

37. Carrano, C.J.; Jordan, M.; Drechsel, H.; Schmid, D.G.; Winkelmann, G. Heterobactins: A new class of siderophores from *Rhodococcus erythropolis* IGTS8 containing both hydroxamate and catecholate donor groups. *Biometals* **2001**, *14*, 119–125. [CrossRef]

38. Han, X.; Liu, Z.; Zhang, Z.; Zhang, X.; Zhu, T.; Gu, Q.; Li, W.; Che, Q.; Li, D. Geranylpyrrol A and Piericidin F from *Streptomyces* sp. CHQ-64 Δ rdmF. *J. Nat. Prod.* **2017**, *80*, 1684–1687. [CrossRef]

39. Nakae, K.; Yoshimoto, Y.; Ueda, M.; Sawa, T.; Takahashi, Y.; Naganawa, H.; Takeuchi, T.; Imoto, M. Migrastatin, a novel 14-membered lactone from *Streptomyces* sp. MK929-43F1. *J. Antibiot.* **2000**, *53*, 1228–1230. [CrossRef] [PubMed]

40. Chen, L.Y.; Wang, X.Q.; Wang, Y.M.; Geng, X.; Xu, X.N.; Su, C.; Yang, Y.L.; Tang, Y.J.; Bai, F.W.; Zhao, X.Q. Genome mining of *Streptomyces xinghaiensis* NRRL B-24674 T for the discovery of the gene cluster involved in anticomplement activities and detection of novel xiamycin analogs. *Appl. Microbiol. Biotechnol.* **2018**, *102*, 9549–9562. [CrossRef] [PubMed]

41. Omura, S.; Matsubara, H.; Nakagawa, A.; Furusaki, A.; Matsumoto, T. X-Ray crystallography of protylonolide and absolute configuration of tylosin. *J. Antibiot.* **1980**, *33*, 915–917. [CrossRef]

42. Gräfe, U.; Schlegel, R.; Stengel, C.; Ihn, W.; Radics, L. Isolation and structure of 26-deoxylaidlomycin, a new polyether antibiotic from *Streptoverticillium olivoreticuli*. *J. Basic Microbiol.* **1989**, *29*, 149–155. [CrossRef]

43. Hirota, A.; Okada, H.; Kanza, T.; Isogai, A.; Hirota, H. Structure elucidation of kaimonolide B, a new plant growth inhibitor macrolide from Streptomyces. *Agric. Biol. Chem.* **1990**, *54*, 2489–2490. [CrossRef]

44. Arakawa, K. Genetic and biochemical analysis of the antibiotic biosynthetic gene clusters on the Streptomyces linear plasmid. *Biosci. Biotechnol. Biochem.* **2014**, *78*, 183–189. [CrossRef]

45. Kusakabe, Y.; Mizuno, T.; Kawabata, S.; Tanji, S.; Seino, A.; Seto, H.; Otake, N. Ferensimycins A and B, two polyether antibiotics. *J. Antibiot.* **1982**, *35*, 1119–1129. [CrossRef]

46. Cho, J.Y.; Williams, P.G.; Kwon, H.C.; Jensen, P.R.; Fenical, W. Lucentamycins A–D, cytotoxic peptides from the marine-derived actinomycete *Nocardiopsis lucentensis*. *J. Nat. Prod.* **2007**, *70*, 1321–1328. [CrossRef] [PubMed]

47. Qian-cutrone, J.; Ueki, T.; Huang, S.; MookhtiaR, K.A.; Ezekiel, R.; Kalinowski, S.S.; Brown, K.S.; Golik, J.; Lowe, S.; Pirnik, D.M.; et al. Glucolipsin A and B, two new glucokinase activators produced by *Streptomyces purpurogeniscleroticus* and *Nocardia vaccinii*. *J. Antibiot.* **1999**, *52*, 245–255. [CrossRef]
48. Howlett, D.R.; George, A.R.; Owen, D.E.; Ward, R.V.; Markwell, R.E. Common structural features determine the effectiveness of carvedilol, daunomycin and rolitetracycline as inhibitors of Alzheimer β-amyloid fibril formation. *Biochem. J.* **1999**, *343*, 419–423. [CrossRef]
49. Dashti, Y.; Grkovic, T.; Abdelmohsen, U.R.; Hentschel, U.; Quinn, R.J. Actinomycete metabolome induction/suppression with N-Acetylglucosamine. *J. Nat. Prod.* **2017**, *80*, 828–836. [CrossRef]
50. Inahashi, Y.; Iwatsuki, M.; Ishiyama, A.; Namatame, M.; Nishihara-Tsukashima, A.; Matsumoto, A.; Hirose, T.; Sunazuka, T.; Yamada, H.; Otoguro, K.; et al. Spoxazomicins A–C, novel antitrypanosomal alkaloids produced by an endophytic actinomycete, *Streptosporangium oxazolinicum* K07-0460 T. *J. Antibiot.* **2011**, *64*, 303–307. [CrossRef]
51. Cheng, K.C.C.; Cao, S.; Raveh, A.; MacArthur, R.; Dranchak, P.; Chlipala, G.; Okoneski, M.T.; Guha, R.; Eastman, R.T.; Yuan, J.; et al. Actinoramide A identified as a potent antimalarial from titration-based screening of marine natural product extracts. *J. Nat. Prod.* **2015**, *78*, 2411–2422. [CrossRef]
52. Elleuch, L.; Shaaban, M.; Smaoui, S.; Mellouli, L.; Karray-Rebai, I.; Fguira, L.F.B.; Shaaban, K.A.; Laatsch, H. Bioactive secondary metabolites from a new terrestrial *Streptomyces* sp. TN262. *Appl. Biochem. Biotechnol.* **2010**, *162*, 579–593. [CrossRef]
53. Giralt, E.; Lo Re, D. The therapeutic potential of migrastatin-core analogs for the treatment of metastatic cancer. *Molecules* **2017**, *22*, 198. [CrossRef]
54. Ujikawa, K.; Vilegas, W.; Vilegas, J.H.; Llabrés, G. Antibiotic 26-deoxylaidlomycin isolated from *Streptomyces* sp. Ar386 from Brazilian soil. *Rev. Latinoam. Microbiol.* **1996**, *38*, 185–191. [PubMed]
55. Le, T.V.; Hanh, T.T.H.; Huong, P.T.T.; Dang, N.H.; Van Thanh, N.; Cuong, N.X.; Nam, N.H.; Thung, D.C.; Van Kiem, P.; Van Minh, C. Anthraquinone and butenolide constituents from the crinoid *Capillaster multiradiatus*. *Chem. Pharm. Bull.* **2018**, *66*, 1023–1026.
56. Kuntsmann, M.P.; Mitscher, L.A. The Structural Characterization of Tetrangomycin and Tetrangulol. *J. Org. Chem.* **1966**, *31*, 2920–2925. [CrossRef] [PubMed]
57. Ukei, M.; Abe, K.; Hanafi, M.; Shibata, K.; Tanaka, T.; Tanaiguchi, M. UK-2 A, B, C and D Novel antifungal antibiotics from *Streptomyces* sp. 517-02. *J. Antibiot.* **1996**, *49*, 639–643. [CrossRef]
58. Guo, Z.K.; Wang, T.; Guo, Y.; Song, Y.C.; Tan, R.X.; Ge, H.M. Cytotoxic angucyclines from *Amycolatopsis* sp HCa1, a rare actinobacteria derived from *Oxya chinensis*. *Planta Med.* **2011**, *77*, 2057. [CrossRef]
59. Hanafi, M.; Shibata, K.; Ueki, M.; Taniguchi, M. UK-2A, B, C and D, Novel Antifungal Antibiotics from *Streptomyces* sp. 517-02. *J. Antibiot.* **1996**, *49*, 1226–1231. [CrossRef] [PubMed]
60. Konopka, J.B. N-acetylglucosamine functions in cell signalling. *Scientifica* **2012**, *2012*, 489208. [CrossRef]
61. Naseem, S.; Parrino, S.M.; Buenten, D.M.; Konopka, J.B. Novel roles for GlcNAc in cell signalling. *Commun. Integr. Biol.* **2012**, *5*, 156–159. [CrossRef] [PubMed]
62. Rigali, S.; Nothaft, H.; Noens, E.E.; Schlicht, M.; Colson, S.; Müller, M.; Joris, B.; Koerten, H.K.; Hopwood, D.A.; Titgemeyer, F.; et al. The sugar phosphotransferase system of Streptomyces coelicolor is regulated by the GntR-family regulator DasR and links N-acetylglucosamine metabolism to the control of development. *Mol. Microbiol.* **2006**, *61*, 1237–1251. [CrossRef] [PubMed]
63. Boonlarppradab, C.; Suriyachadkun, C.; Rachtawee, P.; Choowong, W. Saccharosporones A, B and C, cytotoxic antimalarial angucyclinones from *Saccharopolyspora* sp. BCC 21906. *J. Antibiot.* **2013**, *266*, 305–309. [CrossRef] [PubMed]
64. Carr, G.; Derbyshire, E.R.; Caldera, E.; Currie, C.R.; Clardy, J. Antibiotic and antimalarial quinones from fungus-growing ant-associated *Pseudonocardia* sp. *J. Nat. Prod.* **2012**, *75*, 1806–1809. [CrossRef] [PubMed]
65. Supong, K.; Thawai, C.; Suwanborirux, K.; Choowong, W.; Supothina, S.; Pittayakhajonwut, P. Antimalarial and antitubercular C-glycosylated benz [α] anthraquinones from the marine-derived *Streptomyces* sp. BCC45596. *Phytochem. Lett.* **2012**, *5*, 651–656. [CrossRef]
66. Baragaña, B.; Forte, B.; Choi, R.; Hewitt, S.N.; Bueren-Calabuig, J.A.; Pisco, J.P.; Norcross, N.R. Lysyl-tRNA synthetase as a drug target in malaria and cryptosporidiosis. *Proc. Natl. Acad. Sci. USA* **2019**, *116*, 7015–7020. [CrossRef] [PubMed]
67. Ashelford, K.E.; Chuzhanova, N.A.; Fry, J.C.; Jones, A.J.; Weightman, A.J. At least 1 in 20 16S rRNA sequence records currently held in public repositories is estimated to contain substantial anomalies. *Appl. Environ. Microbiol.* **2005**, *71*, 7724–7736. [CrossRef] [PubMed]
68. Pruesse, E.; Peplies, J.; Glockner, F.O. SINA: Accurate high-throughput multiple sequence alignment of ribosomal RNA genes. *Bioinformatics* **2012**, *28*, 1823–1829. [CrossRef]
69. Letunic, I.; Bork, P. Interactive Tree of Life v2: Online annotation and display of phylogenetic trees made easy. *Nucleic Acids Res.* **2011**, *39*, W475–W478. [CrossRef]
70. Ngwa, C.J.; Kiesow, M.J.; Papst, O.; Orchard, L.M.; Filarsky, M.; Rosinski, A.N.; Voss, T.S.; Llinás, M.; Pradel, G. Transcriptional profiling defines histone acetylation as a regulator of gene expression during human-to-mosquito transmission of the malaria parasite *Plasmodium falciparum*. *Front. Cell. Infect. Microbiol.* **2017**, *7*, 320. [CrossRef]
71. Basova, S.; Wilke, N.; Koch, J.C.; Prokop, A.; Berkessel, A.; Pradel, G.; Ngwa, C.J. Organoarsenic Compounds with in vitro Activity against the Malaria Parasite Plasmodium falciparum. *Biomedicines* **2020**, *8*, 260. [CrossRef] [PubMed]

72. Sayed, A.M.; Alhadrami, H.A.; El-Gendy, A.O.; Shamikh, Y.I.; Belbahri, L.; Hassan, H.M.; Rateb, M.E. Microbial natural products as potential inhibitors of SARS-CoV-2 main protease (Mpro). *Microorganisms* **2020**, *8*, 970. [CrossRef] [PubMed]
73. Sayed, A.M.; Alhadrami, H.A.; El-Hawary, S.S.; Mohammed, R.; Hassan, H.M.; Rateb, M.E.; Bakeer, W. Discovery of two brominated oxindole alkaloids as Staphylococcal DNA gyrase and pyruvate kinase inhibitors via inverse virtual screening. *Microorganisms* **2020**, *8*, 293. [CrossRef] [PubMed]

 *marine drugs*

Article

Antimicrobial Chlorinated 3-Phenylpropanoic Acid Derivatives from the Red Sea Marine Actinomycete *Streptomyces coelicolor* LY001

Lamiaa A. Shaala [1,2,3,*], **Diaa T. A. Youssef** [4,5,*], **Torki A. Alzughaibi** [2,6] **and Sameh S. Elhady** [4]

1 Natural Products Unit, King Fahd Medical Research Center, King Abdulaziz University, Jeddah 21589, Saudi Arabia
2 Department of Medical Laboratory Sciences, Faculty of Applied Medical Sciences, King Abdulaziz University, Jeddah 21589, Saudi Arabia; taalzughaibi@kau.edu.sa
3 Suez Canal University Hospital, Suez Canal University, Ismailia 41522, Egypt
4 Department of Natural Products, Faculty of Pharmacy, King Abdulaziz University, Jeddah 21589, Saudi Arabia; ssahmed@kau.edu.sa
5 Department of Pharmacognosy, Faculty of Pharmacy, Suez Canal University, Ismailia 41522, Egypt
6 King Fahd Medical Research Center, King Abdulaziz University, Jeddah 21589, Saudi Arabia
* Correspondence: Lshalla@kau.edu.sa (L.A.S.); dyoussef@kau.edu.sa (D.T.A.Y.);
  Tel.: +966-548-751-044 (L.A.S.); +966-548-535-344 (D.T.A.Y.)

Received: 22 July 2020; Accepted: 25 August 2020; Published: 27 August 2020

Abstract: The actinomycete strain *Streptomyces coelicolor* LY001 was purified from the sponge *Callyspongia siphonella*. Fractionation of the antimicrobial extract of the culture of the actinomycete afforded three new natural chlorinated derivatives of 3-phenylpropanoic acid, 3-(3,5-dichloro-4-hydroxyphenyl)propanoic acid (**1**), 3-(3,5-dichloro-4-hydroxyphenyl)propanoic acid methyl ester (**2**), and 3-(3-chloro-4-hydroxyphenyl)propanoic acid (**3**), together with 3-phenylpropanoic acid (**4**), *E*-cinnamic acid (**5**), and the diketopiperazine alkaloids cyclo(L-Phe-*trans*-4-OH-L-Pro) (**6**) and cyclo(L-Phe-*cis*-4-OH-D-Pro) (**7**) were isolated. Interpretation of nuclear magnetic resonance (NMR) and high-resolution electrospray ionization mass spectrometry (HRESIMS) data of **1–7** supported their assignments. Compounds **1–3** are first candidates of the natural chlorinated phenylpropanoic acid derivatives. The production of the chlorinated derivatives of 3-phenylpropionic acid (**1–3**) by *S. coelicolor* provides insight into the biosynthetic capabilities of the marine-derived actinomycetes. Compounds **1–3** demonstrated significant and selective activities towards *Escherichia. coli* and *Staphylococcus aureus*, while *Candida albicans* displayed more sensitivity towards compounds **6** and **7**, suggesting a selectivity effect of these compounds against *C. albicans*.

Keywords: Red Sea sponges; marine actinomycetes; *Streptomyces coelicolor* LY001; halogenated 3-phenylpropanoic acid derivatives; diketopiperazine alkaloids; structural determinations; antimicrobial activities

---

1. Introduction

The marine actinomycetes represent a vital source of biologically active secondary metabolites and a promising future source for drug discovery. It is well known that marine tunicates and sponges are highly associated with symbiotic microbes [1–3]. There are very few reports about investigation of Red Sea actinomycetes for their chemical diversity and biomedical importance [4].

Streptomycetes represent a group within the actinomycetes with an economical importance and represent a vigorous source of different bioactive secondary metabolites [5]. More than 75% of the marketed antibiotics, commercially available compounds and several agrochemicals are synthesized by streptomycetes [6–8]. In 1990, the genus *Streptomyces* alone represented the main source of almost 60% of the antibiotics and most of the agrochemicals [9]. Streptomycetes produce an array of diverse and bioactive compounds with antimicrobial [5,9–11], anticancer [5,12,13], insecticidal and antiparasitic [14], anti-inflammatory [15], anti-fouling [16], antiviral and anti-infective [17,18] properties. Moreover, the genus *Streptomyces* is considered as a candidate of industrial importance [19–21], and a producer of secondary metabolites with herbicidal activity and which promote plant growth [22], vitamins [23], ribonucleases [11,13,24–27], and enzyme inhibitors [13]. These features make these microorganisms an ideal and favorite research project for academia and industry [9]. Recent advances in marine microbiology, including their purification and identification of bacteria that produce rich arrays of bioactive natural products, provide a strong motivation to explore aggressively their potential as sources of novel pharmaceutical agents [28–30]. Likewise, new methodology for rapid and affordable sequencing of microbial chromosomes, bioinformatic and metabolic profiling enables information on the predicted secondary metabolic diversity contained within a bacterial genome to be rapidly evaluated and manipulated [31–33].

Due to the fast growing number of pathogenic microbes, viral infections and cancer cells that have resistance towards current therapies, drug leads, and new chemical entities for the enhancement of new drugs are in high demand. Tuberculosis, malaria, and *Staphylococcus aureus* infections are among just a few diseases that have become difficult to treat with antibiotics [34]. There is, therefore, a pressing need for the development of new methodologies to provide new drugs/drug leads for the future. Marine microorganisms, such as actinomycetes that exist in association with marine invertebrates and algae, produce novel chemical entities with potential pharmaceutical significance. The importance of natural products, however, extends far beyond that of drug discovery to that of addressing fundamental biological questions, taking full advantage of their structural complexity and functional diversity to probe biological function.

The decreasing number of FDA approved therapeutics, and the low number of drugs in the pipeline that have arisen from synthetic combinatorial libraries, has spurred renewed interest in natural products research with the aim of identifying new structural scaffolds. Marine-derived actinomycetes represent a promising and vigorous source of ubiquitous and diverse and bioactive secondary metabolites that can be obtained in large-scale cultures and developed as drug leads. Overall, there are relatively few reports of marine microbial cultivation from the highly biodiverse Red Sea. The results from this study confirm that Red Sea marine actinomycetes represent a promising source of new drug leads with antibiotic potential.

As a part of our endeavor to purify and characterize bioactive marine microbial candidates [35,36], the antimicrobial fractions of extract of the culture of *Streptomyces coelicolor* LY001 was investigated. Three chlorinated derivatives of 3-phenylpropanoic acid, including 3-(3,5-dichloro-4-hydroxyphenyl)propanoic acid (**1**), 3-(3,5-dichloro-4-hydroxyphenyl)propanoic acid methyl ester (**2**), and 3-(3-chloro-4-hydroxyphenyl)propanoic acid (**3**), along with 3-phenylpropanoic acid (**4**) [37], *E*-cinnamic acid (**5**) and the diketopiperazine alkaloids cyclo(L-Phe-*trans*-4-OH-L-Pro) (**6**) [38,39], and cyclo(L-Phe-*cis*-4-OH-D-Pro) (**7**) [40,41] were isolated and identified. Compounds **1**–**7** were determined by assignments of their NMR and HRESIMS data. Herein, the isolation, structure assignments, and the antimicrobial activities of **1**–**7** are presented.

2. Results and Discussion

2.1. Isolation of the Actinomycete, Streptomyces coelicolor LY001

The marine-derived actinomycete LY001 (Figure 1) was isolated from the internal tissues of the sponge *Callyspongia siphonella* (Figure 1). The resulting sequence of the actinomycete strain was searched for homology with Basic Local Alignment Search Tool (BLAST) in the GenBank. The alignment with reported sequences in the GenBank showed that the LY001 isolate belongs to the genus *Streptomyces* and displayed 100% similarity with the strain *Streptomyces coelicolor* AB588124.

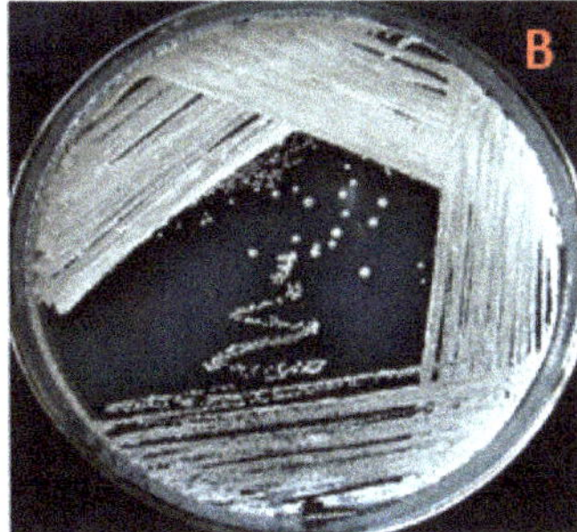

Figure 1. The Red Sea sponge *Callyspongia siphonella* (**A**) and the actinomycete *Streptomyces coelicolor* LY001 (**B**).

2.2. Structure Elucidation of the Compounds

The structural assignment of compound **1** (Figure 2) was supported by interpretation of its NMR spectra (Figures S1–S5). The molecular formula of $C_9H_8Cl_2O_3$ was suggested for **1** as supported by HRESIMS (234.9932, $C_9H_9Cl_2O_3$, $[M + H]^+$) (Figure S6), suggesting five degrees of unsaturation. Its ^{13}C NMR spectrum along with the heteronuclear single-quantum correlation spectroscopy (HSQC) experiment showed seven signals equivalent for nine carbons including the equivalent methines (C-2 and C-6), two methylenes (C-7 and C-8) and five quaternary carbons, including two chemically equivalent carbons (C-3 and C-5) (Table 1). Its ^1H NMR demonstrated a 2-proton singlet at δ_H 7.12 for the chemically equivalent protons H-2 and H-6 (Table 1). This two-proton singlet at δ_H 7.12 was correlated to the signal at δ_C 128.1 (CH, C-2/C-6) in the HSQC experiment. In the ^1H-^1H correlation spectroscopy (COSY) spectrum, vicinal couplings between the methylene protons at δ_H 2.86 (2H, H_2-7) and 2.64 (2H, H_2-8) was observed. Further, the protons of the methylenes at δ_H 2.86 (H_2-7) and 2.64 (H_2-8) were correlated, in the HSQC experiment, to the ^{13}C NMR signals at δ_C 39.3 (CH_2, C-7) and 34.5 (CH_2, C-8), respectively. In addition, the existence of the quaternary carbons at δ_C 133.6 (qC, C-1), 122.1 (qC, C-3, C-5), and 146.0 (qC, C-4) together with δ_C 128.1 (2 × CH, C-2/C-6) suggested a 1,3,4,5-tetrasubstituted benzene moiety with OH at C-4 and symmetrical substitutions with chlorine atoms at C-3 and C-5.

Table 1. NMR data of **1–3** (CDCl$_3$, ^1H at 850 MHz, ^{13}C at 213 Hz).

Position	1		2		3	
	δ_C (mult.) [a]	δ_H (mult., J in Hz)	δ_C (mult.) [a]	δ_H (mult., J in Hz)	δ_C (mult.) [a]	δ_H (mult., J in Hz)
1	133.6, qC		133.8, qC		133.5, qC	
2	128.1, CH	7.12 (s)	128.1, CH	7.10 (s)	128.6, CH	7.17 (d, 2.5)
3	122.1, qC		121.0, qC		120.8, qC	
4	146.9, qC		146.6, qC		151.7, qC	
5	122.1, qC		121.0, qC		116.1, CH	6.93 (d, 8.5)
6	128.1, CH	7.12 (s)	128.1, CH	7.10 (s)	128.4, CH	7.02 (dd, 8.5, 2.5)
7	29.3, CH$_2$	2.86 (t, 7.6)	29.8, CH$_2$	2.84 (t, 7.6)	29.5, CH$_2$	2.86 (t, 7.6)
8	34.5, CH$_2$	2.64 (t, 7.6)	35.3, CH$_2$	2.59 (t, 7.6)	34.8, CH$_2$	2.65 (t, 7.6)
9	173.8, qC		173.3, qC		173.9, qC	
10			51.9, CH$_3$	3.67 (s)		

[a] Multiplicities of the ^{13}C NMR signals were assigned from HSQC experiment.

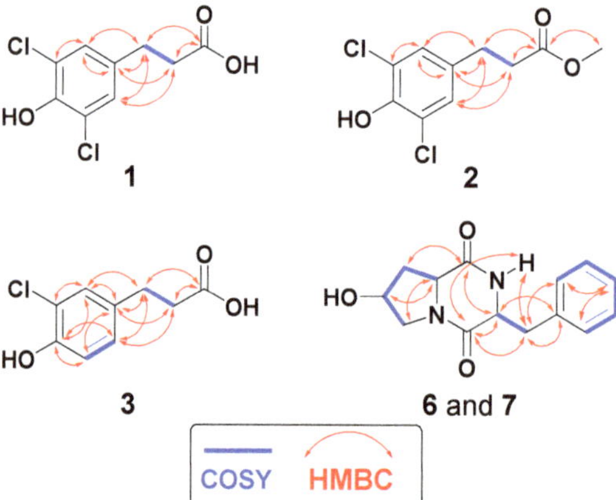

Figure 2. Structures of **1–7**.

^{1}H-^{13}C Heteronuclear multiple bond correlation spectroscopy (HMBC) experiment supported the substitution on the phenyl moiety. HMBC of H-2,6/C-1 (qC, δ_C 133.6), H-2,6/C-3,5 (qC, δ_C 122.1), and H-2,6/C-4 (qC, δ_C 146.9) supported this substitution (Figure 3). Thus, 3,5-dichloro-4-hydroxyphenyl moiety was assigned as part A of the molecule (Figure 2). Furthermore, the ^{1}H chemical shifts values of H-2 and H-6, and ^{13}C values of C-1–C-6 in **1** are similar with those of (*S*)-2-amino-3-(3,5-dichloro-4-hydroxyphenyl)-propanoic acid [42]. The substituent at C-1 of the phenyl moiety was assigned as 3-substituted propanoic acid (part B) as supported from the signals at $\delta_{H/C}$ 2.86 (2H)/29.3 (H$_2$-7/C-7), 2.64 (2H)/34.5 (H$_2$-8/C-8), and δ_C 173.8 (qC, C-9) (Table 1). The HMBC from the protons at H$_2$-7 and H$_2$-8 to C-9 (δ_C 173.8) (Figure 3) supported this assignment. Finally, the connection between the two parts (A and B) of **1** was proved by cross-peaks in the HMBC of H$_2$-7/C-1, H$_2$-7/C-2,6, H$_2$-8/C-1, and H-2,6/C-7 (Figure 3). Thus, compound **1** was assigned as 3-(3,5-dichloro-4-hydroxyphenyl)propanoic acid and this is the first report about its natural occurrence.

Figure 3. ^{1}H-^{1}H COSY and ^{1}H-^{13}C HMBC of **1-3**, **6**, and **7**.

Interpretation of the one- and two-dimensional NMR data (Figures S7–S11) supported the structural determination of **2** (Figure 2). It possesses molecular formula C$_{10}$H$_{10}$Cl$_2$O$_3$ as supported by HRESIMS (249.0088, C$_{10}$H$_{11}$Cl$_2$O$_3$, [M + H]$^+$) (Figure S12), being 15 mass units larger than **1**, proving the existence of an additional CH_3 in **2**. The structure of **2** was assigned by interpretation of its NMR spectra. The NMR data of **2** are identical with those of **1** (Table 1), suggesting similar structures. Furthermore, the existence of an additional three-proton singlet at δ_H 3.67 correlated to the signal at δ_C 51.9 (CH_3, C-10) in the HSQC, which supports the existence of a terminal methoxyl group in **2**. The HMBC from H_3-10 (δ_H 3.67) to C-9 (δ_C 173.3) supported the existence of a methyl ester group in **2** instead of a free carboxylic acid in **1**.

Further, the assignment of the substituents on the aromatic moiety was supported by HMBC (Figure 3) as previously discussed under compound **1**. Similarly, the ^{1}H chemical shifts of H-2 and H-6, and ^{13}C chemical shifts values of C-1–C-6 in **2** are similar to those of (*S*)-2-amino-3-(3,5-dichloro-4-hydroxyphenyl)-propanoic acid [42]. Accordingly, compound **2** was assigned as 3-(3,5-dichloro-4-hydroxyphenyl)propanoic acid methyl ester and this is its first natural occurrence.

The structure determination of compound **3** (Figure 2) was assigned by examination of its NMR spectroscopic data (Figures S13–S17). Compound **3** with molecular formula of $C_9H_9ClO_3$ (201.0321, $C_9H_{10}ClO_3$, [M + H]$^+$) (Figure S18), being 34 units less than **1**, suggests the presence of only one chlorine atom in **3**. Its ^{13}C NMR spectrum and HSQC exhibited nine signals for three aromatic methines (C-2, C-5, and C-6), two methylenes (C-7 and C-8), and four quaternary carbons (Table 1). Its ^{1}H NMR spectrum showed two spin-spin coupling systems. The first one includes the aromatic ABX coupling system at δ_H 7.17 (H-2, d, J = 2.5 Hz), 6.93 (H-5, d, J = 8.5 Hz), and 7.02 (H-6, dd, J = 8.5, 2.5 Hz) (Table 1). These protons are correlated in the HSQC experiment to the signals at δ_C 128.6 (C-2), 116.1 (C-5), and 128.4 (C-6), respectively. This system supported the existence of a 1,3,4-trisubstituted aromatic moiety in **3**. The second spin coupling system includes the vicinal coupling between the methylenes at δ_H 2.86 (H_2-7, t, J = 7.6 Hz) and 2.65 (H_2-8, t, J = 7.6 Hz). In the HMBC, H_2-7 (δ_H 2.86) and H_2-8 ((δ_H 2.65) displayed correlation to the signal at δ_C 173.9 (C-9) (Figure 3), supporting the existence of a 3-substituted propanoic acid moiety as in **1**. Again, the placement of the side chain (3-substituted propanoic acid) at C-1 was secured from HMBC of H-2/C-7, H-6/C-7, H_2-7/C-1, and H_2-8/C-1 (Figure 3). Finally, HMBC of H-2/C-3 (qC, δ_C 120.8), H-2/C-4 (qC, δ_C 151.7), H-5/C-4, H-5/C-3, H-5/C-1 (qC, δ_C 133.5), H_2-7/C-9 (qC, δ_C 173.9), and H_2-8/C-9 (Figure 3) supported the substitution on the benzene ring in **3**. Again, the chemical shifts of the ^{1}H (H-2, H-5 and H-6) and ^{13}C (C-1–C-6) NMR signals of the 3,5-dichloro-4-hydroxy phenyl moiety in compound **3** are similar with reported data for (*S*)-3-(3-chloro-4-hydroxyphenyl)-2-(dimethylamino)propanoic acid [42]. Thus, **3** was assigned as 3-(3-chloro-4-hydroxyphenyl)propanoic acid and this is the first report about its natural occurrence.

Interpretation of the NMR (Figures S19–S23) and MS spectra of **4** (Figure 2) supported its structure. The NMR are identical with those reported for 3-phenylpropanoic acid from a terrestrial *Streptomyces* strain [37]. Accordingly, **4** was assigned as 3-phenylpropanoic acid and this is its first occurrence from a marine *Streptomyces*.

The structure of **5** (Figure 2) was assigned, from the NMR (Figures S24 and S25) and MS spectra, as *E*-cinnamic acid. Its NMR spectra exhibited signals for a monosubstituted benzene ring. The J value ($J_{7,8}$ = 16.1 Hz) supported *E* configuration in **5**.

The structures of **6** and **7** (Figure 2) were identified from interpretation of the corresponding NMR (Figures S26–S30 for **6** and Figures S31–S35 for **7**) and MS spectra. Both compounds (Figure 2) displayed the same molecular formula of ($C_{14}H_{16}N_2O_3$). The diketopiperazine nature of **6** and **7** are obvious from the corresponding ^{1}H and ^{13}C resonating signals (Table 2). Those include the resonances of two amidic carbonyls (C-2 and C-5) as well as the resonances for the methine signals at C-3 and C-6 (Table 2). The remaining of the signals are characteristic for the presence of a hydroxylated proline and phenylalanine moieties in **6** and **7** [38–41]. The NMR signals at $\delta_{H/C}$ 4.60/68.4 (in **6**) and $\delta_{H/C}$ 4.40/68.1 (in **7**) are characteristic for the hydroxylated C-8 supporting the hydroxylation of the proline moieties in both compounds [38–41]. The NMR data of **6** and **7** (Table 2) are similar to those of cyclo(L-Phe-*trans*-4-OH-L-Pro) [38,39] and cyclo(L-Phe-*cis*-4-OH-D-Pro) [40,41], respectively. Furthermore, the 2D experiments (HSQC, COSY, HMBC) (Figure 3) supported the assignment of all signals of the compounds (Table 2 and Figure 3). Accordingly, compounds **6** and **7** were assigned as cyclo(L-Phe-*trans*-4-OH-L-Pro) and cyclo(L-Phe-*cis*-4-OH-D-Pro).

Table 2. NMR data of **6** and **7** (CDCl₃).

Position	6 (¹H at 600 MHz, ¹³C at 150 MHz)		7 (¹H at 850 MHz, ¹³C at 213 Hz)	
	δ$_C$ (mult.) [a]	δ$_H$ (mult., *J* in Hz)	δ$_C$ (mult.) [a]	δ$_H$ (mult., *J* in Hz)
2	165.0, qC		165.2, qC	
3	56.1, CH	4.33 (dd, 10.8, 3.0)	59.1, CH	4.23 (tt, 4.2)
4 (N*H*)		5.60 (brs)		5.80 (brs)
5	169.5, qC		168.4, qC	
6	57.3, CH	4.46 (dd, 10.8, 6.0)	55.8, CH	3.17 (t, 8.5)
7a	37.8, CH₂	2.37 (dd, 13.8, 6.0)	38.0, CH₂	2.37 (ddd, 13.7, 8.5, 5.1)
7b		2.06 (ddd, 13.8, 11.4, 4.2)		2.20 (m)
8	68.4, CH	4.60 (t, 4.2)	68.1, CH	4.40 (quin, 4.0)
9a	54.4, CH₂	3.80 (dd, 13.2, 4.2)	53.4, CH₂	3.82 (dd, 12.7, 2.5)
9b		3.58 (d, 13.2)		3.35 (dd, 12.7, 5.1)
10a	36.6, CH₂	3.63 (dd, 14.5, 4.2)	40.8, CH₂	3.13 (dd, 14.0, 6.8)
10b		2.77 (dd, 15.5, 10.8)		3.10 (dd, 14.0, 4.2)
11	135.7, qC		135.2, qC	
12	129.0, CH	7.22 (d, 7.5)	129.7, CH	7.21 (d, 7.3)
13	129.3, CH	7.36 (t, 7.5)	128.9, CH	7.31 (t, 7.3)
14	127.6, CH	7.29 (t, 7.5)	127.6, CH	7.30 (t, 7.3)
15	129.3, CH	7.36 (t, 7.5)	128.9, CH	7.31 (t, 7.3)
16	129.0, CH	7.22 (d, 7.5)	129.7, CH	7.21 (d, 7.3)

[a] Multiplicities of the ¹³C NMR signals were assigned from HSQC experiment.

The derivatives of 3-phenylpropanoic acid are rarely reported from microbial organisms. We believe that there is only one report about the occurrence of 3-phenylpropanoic acid from a terrestrial *Streptomyces* strain [37]. Further, it is worth pointing out that this is the first report about the natural existence of chlorinated derivatives of 3-phenylpropanoic acid. These results give an understanding about the biosynthetic potential and structural diversity of the cultured marine-derived microbes and the future application of these metabolites in drug discovery.

Phenylpropanoids area ubiquitous group of organic compounds including *flavonoids*, coumarins, phenolic acids, stilbenes, and lignins. They originate from phenylalanine and tyrosine [43]. The name phenylpropanoid is derived from a six-carbon forming a phenyl moiety connected to a three-carbon propene unit. The coumaric acid represents the key biosynthetic intermediate in the biosynthesis of all phenylpropanoids. Biosynthetically, a variety of natural products including flavonoids, phenylpropanoids, isoflavonoids, catechins, coumarins, stilbenes, aurones, and lignols originates from 4-coumaroyl-CoA [44], which is originated from cinnamic acid [44].

Three pathogens were used to determine the antimicrobial effects and the minimum inhibitory concentration (MIC) values of **1–7**. Compound **1** showed the greatest inhibitory effects towards *S. aureus* and *E. coli* with inhibition zones of 17 and 23 mm (Table 3). In addition, compounds **2** and **3** were less active than **1** against these pathogens with inhibition zones of 12–20 mm (Table 3). Compound **4** displayed the lowest activity against these pathogens with inhibition zones of 15 and 11 mm. These findings suggest the importance of the substitution with a "3,5-dichloro-4-hydroxy" moiety on the phenyl moiety as well as the presence of a free terminal carboxylic acid moiety for a maximum antibacterial activity (as in **1**). On the other hand, **1–4** showed modest effect towards *C. albicans* (ATCC 14053) with 6–9 mm inhibition zone (Table 3), suggesting selective antibacterial effects of these compounds against *E. coli* and *S. aureus*. Finally, the diketopiperazine alkaloids **6** and **7** displayed better activities towards *C. albicans* with 12–14 mm inhibition zones, while they were less active towards *E. coli* and *S. aureus* with 7–11 mm inhibition zones, suggesting their selective antifungal activity against *C. albicans*.

Table 3. Antimicrobial effects of **1–7**.

Compound	*E. coli*		*S. aureus*		*C. albicans*	
	Inhibition Zone (mm)	MIC (µg/mL)	Inhibition Zone (mm)	MIC (µg/mL)	Inhibition Zone (mm)	MIC (µg/mL)
1	23	16	17	32	9	125
2	20	32	12	64	7	250
3	18	32	15	32	7	250
4	15	64	11	125	6	250
5	NT	NT	NT	NT	NT	NT
6	7	250	9	250	12	32
7	11	125	7	250	14	32
Ciprofloxacin [a]	30	0.25	22	0.5	NT	NT
Ketoconazole [b]	NT	NT	NT	NT	30	0.5

[a] positive antibacterial drug; [b] positive antifungal drug.

To determine the MIC values of the compounds, a microdilution method was carried out (Table 3). Compound **1** displayed the highest activity towards *E. coli* with an MIC of 16 µg/mL, while compounds **6** and **7** displayed the highest antifungal activities with an MIC of 32 µg/mL towards *C. albicans*. On the other hand, compounds **2** and **3** displayed lower activities with MIC values of 32–64 µg/mL towards *E. coli* and *S. aureus*. Other compounds were weakly active with MIC values of 64–250 µg/mL.

3. Materials and Methods

3.1. General Experimental Procedures

Optical rotations, ultraviolet (UV), infra-red (IR), NMR, and HRESIMS were acquired as previously reported [35,36]. Fractionation of the extracts and successive fractions were performed on SiO_2 and Sephadex LH-20. The purification of the compounds was carried out on an analytical Shim-Pack C18 (250 × 4.6 mm, Shimadzu, Kyoto, Japan).

3.2. The Host Organism, C. siphonella

The marine sponge *C. siphonella* was harvested in May 2016 using scuba at a depth up to 20 m off Jizan at the Saudi Red Sea. The pink tubular sponge is dichotomously divided with a smooth thin-walled surface. It possesses a soft compressible consistency, which is difficult to tear. The voucher specimen measures up to 10 cm, while the branching tubes measure up to 5 cm in height and up to 2.5 cm in diameter. A comprehensive description of the sponge and the specimen's codes are previously reported [45,46].

3.3. Isolation of the Actinomycete Streptomyces coelicolor LY001

After surface sterilization, about 1 cm^3 of the internal tissue of the sponge was finely mixed in sterile seawater (10 mL), diluted, and spread on International Streptomyces Project-2 (ISP2) medium. The medium was amended with 3% NaCl. Afterwards, cultured plates were incubated at 30 °C and checked after actinomycetes growth regularly. The actinomycete LY001 was obtained in a pure state after several purification steps.

3.4. Characterization of the Actinomycete, Streptomyces coelicolor

The LY001 strain was identified by analysis of its 16S rRNA sequence. DNA preparation was used for 16S rRNA gene PCR amplification using 27f and 1492r primers. The reaction mixture composed of 50 µL included 1000 ng of gDNA, primers (each 20 pmol), and GoTaq Master Mixes (25 µL). The polymerase chain reaction (PCR) thermocycler initiated with denaturation at 95 °C (2 min), 30 cycles at 95 °C (30 s for denaturation), 30 s at 58 °C (for annealing), 60 s at 72 °C (for extension), and at 72 °C for 5 min for final completion of DNA extension. Purification of PCR products was accomplished on Agarose Gel DNA Purification Kit (Biocompare, South San Francisco, CA, USA) as supported by the supplier. 16S rRNA sequence displayed 100% similarity with *Streptomyces coelicolor* (Accession No. AB588124). The sequence of the Red Sea *Streptomyces coelicolor* LY001 was placed in the NCBI GenBank under the Accession Number MN883509 on 30 December 2019 (http://getentry.ddbj.nig.ac.jp/).

3.5. Large-Scale Culture of Streptomyces coelicolor

Spores of *Streptomyces coelicolor* were cultured in 2.0 L flasks, each containing ISP2 media (500 mL) [47] including 10 g of malt extract, 4.0 g of yeast extract, 4.0 g of dextrose, and 3% NaCl (*w/v*) in 1 L distilled water at pH of 7. Incubation of the culture was accomplished by shaking at 180 rpm at 28 °C for 14 days. The combined culture broth (10 L) was shaken against EtOAC three times (each 3 L). The resulted EtOAc extracts dried to give 1.3 g.

3.6. Purification of 1–7

The dried extract (1.3 g) was partitioned on C18 (ODS) silica (Sigma Aldrich, St. Louis, MO, USA) using vacuum liquid chromatography (VLC)column using H_2O-MeOH gradients to give 10 fractions (Fractions A–J). The antimicrobial fractions C (180 mg) and D (118 mg) were separately fractionated on Sephadex LH-20 using MeOH-CH_2Cl_2 (1:1) mixture to give five subfractions, each. Subfraction C4 (65 mg) was purified on an ODS HPLC column using 30% CH_3CN to afford compounds **5** (9.1 mg), **6** (3.7 mg), and **7** (2.6 mg). Subfraction D4 (54 mg) was purified on an ODS HPLC using 45% CH_3CN to afford compounds **1** (6.2 mg), **2** (3.6 mg), **3** (3.1 mg), and **4** (4.2 mg).

Spectral Data of **1–7**

3-(3,5-Dichloro-4-hydroxyphenyl)propanoic acid (**1**). Colorless solid; UV (MeOH) λ_{max} (log ε): 227 (4.15), 312 (4.17) nm; IR (film) v_{max} 3340, 3029, 1700, 1301, 1219, 935 cm^{-1}; HRESIMS *m/z* 234.9932 (calcd for $C_9H_9Cl_2O_3$, [M + H]$^+$, 234.9929), NMR: Table 1.

3-(3,5-Dichloro-4-hydroxyphenyl)propanoic acid methyl ester (**2**). Colorless solid; UV (MeOH) λ_{max} (log ε): 229 (4.15), 315 (4.17) nm; IR (film) v_{max} 3335, 3030, 1665, 1305, 1219, 937 cm^{-1}; HRESIMS *m/z* 249.0088 (calcd for $C_{10}H_{11}Cl_2O_3$, [M + H]$^+$, 249.0085); NMR: Table 1.

3-(3-Chloro-4-hydroxyphenyl)propanoic acid (**3**). Colorless solid; UV (MeOH) λ_{max} (log ε): 225 (3.95), 307 (4.00) nm; IR (film) v_{max} 3341, 3030, 1702, 1303, 1221, 936 cm^{-1}; HRESIMS *m/z* 201.0321 (calcd for $C_9H_{10}ClO_3$, [M + H]$^+$, 201.0318), NMR: Table 1.

3-Phenylpropanoic acid (**4**). Colorless solid; ESIMS *m/z* 151.07 ($C_9H_{11}O_2$, [M + H]$^+$), NMR data: ^{1}H NMR (850 MHz, CDCl$_3$) δ_H: 7.22 (2H, m, H-2,6), 7.30 (2H, m, H-3,5), 7.21 (1H, m, H-4), 2.97 (t, *J* = 7.5 H$_2$-7), 2.69 (t, *J* = 7.5 H$_2$-8); ^{13}C NMR (312 MHz, CDCl$_3$) δ_C: 140.1 (qC, C-1), 128.5 (CH, C-2,6), 128.2 (CH, C-3,5), 126.4 (CH, C-4), 29.7 (CH$_2$, C-7), 34.7 (CH$_2$, C-8), 171.6 (qC, C-9).

E-Cinnamic acid (**5**). Colorless solid; ESIMS *m/z* 149.06 ($C_9H_9O_2$, [M + H]$^+$); NMR data: ^{1}H NMR (850 MHz, CDCl$_3$) δ_H: 7.55 (2H, m, H-2,6), 7.41 (2H, m, H-3,5), 7.40 (1H, m, H-4), 7.79 (1H, d, *J* = 16.1 Hz, H-7), 6.45 (1H, d, *J* = 16.1 Hz, H-8); ^{13}C NMR (312 MHz, CDCl$_3$) δ_C 134.4 (C-1), 128.1 (C-2,6), 128.9 (C-3,5), 130.8 (C-4), 147.2 (C-7), 117.0 (C-8), 171.6 (C-9).

Cyclo(L-Phe-*trans*-4-OH-L-Pro) (**6**). Colorless solid; [α]$_D$ −55° (c 0.10, MeOH); UV (MeOH) λ_{max} (log ε): 230 (4.10), 305 (4.07) nm; IR (film) v_{max} 3451, 1663, 1629 cm^{-1}; HRESIMS *m/z* 261.1241 (calcd for $C_{14}H_{17}N_2O_3$, [M + H]$^+$, 261.1239), NMR: Table 2.

Cyclo(L-Phe-cis-4-OH-D-Pro) (**7**). Colorless solid; $[\alpha]_D$ +38° (c 0.10, MeOH); UV (MeOH) λ_{max} (log ε): 230 (4.10), 305 (4.07) nm; IR (film) ν_{max} 3450, 1662, 1629 cm^{-1}; HRESIMS m/z 261.1241 (calcd for $C_{14}H_{17}N_2O_3$, [M + H]$^+$, 261.1239), NMR: Table 2.

3.7. Antimicrobial Evaluation of Compounds **1–7**

3.7.1. Disc Diffusion Assay

Using the disc diffusion assay, the antimicrobial effects of **1–7** were evaluated at 100 µg/disc against several pathogenic microbes including *E. coli* (ATCC 25922), *C. albicans* (ATCC 14053), and *S. aureus* (ATCC 25923) and as previously described [48–50]. Ciprofloxacin (5.0 µg/disc) and ketoconazole (50 µg/disc) were used as positive antibiotics.

3.7.2. Determination of the MIC of **1–7**

The MIC values of the compounds was evaluated using a macrodilution method [51]. Briefly, MeOH was used to dissolve the compounds at a final concentration of 2000 µg/mL, while distilled water was used to dissolve ciprofloxacin and ketoconazole at final concentrations of 100 µg/mL. All solutions were sterilized using syringe filters (0.2 µm). A two-fold serial dilution of the solutions was used in Mueller Hinton Broth (*MHB*) to afford concentrations between 1.0 and 1000 µg/mL for the compounds and between 0.125 and 64 µg/mL for ciprofloxacin and ketoconazole. From the 10^6 colony-forming units (CFU)/mL microbial suspensions, 500 µL were added in sterile tubes giving inoculua of 5×10^5 CFU/mL. Additional 100 µL of each stock solution of the compounds and antibiotics were added into the tubes. A control tube, which contains only the test microorganisms and methanol was prepared. The MeOH displayed no antimicrobial effect. Incubation of the tubes was accomplished at 37 °C for 48 h. The lowest concentrations of the compounds/antibiotics, which show no microbial growth were considered as MIC.

4. Conclusions

Purification of the antimicrobial fractions of the culture of the *Streptomyces coelicolor* LY001 afforded three chlorinated 3-phenylpropanoic acid derivatives including 3-(3,5-dichloro-4-hydroxyphenyl) propanoic acid (**1**), 3-(3,5-dichloro-4-hydroxyphenyl)propanoic acid methyl ester (**2**), and 3-(3-chloro-4-hydroxyphenyl)propanoic acid (**3**) along with 3-phenylpropanoic acid (**4**), E-cinnamic acid (**5**), and the dipeptides cyclo(L-Phe-*trans*-4-OH-L-Pro (**6**) and cyclo(L-Phe-*cis*-4-OH-D-Pro) (**7**). Structures of **1–7** were determined from interpretation of their NMR and HRESIMS spectroscopic data. The chlorinated 3-phenylpropanoic acid derivatives (**1–3**) showed selective and significant antimicrobial activities against *S. aureus* and *E. coli* and displayed modest effects towards *C. albicans*. On the other hand, the diketopiperazine alkaloids **6** and **7** were more active against and selective against *C. albicans* and less active than *E. coli* and *S. aureus*. Interestingly, this is the first report about natural occurrence of chlorinated 3-phenypropionic acid derivatives from a cultured marine-derived *Streptomyces*. These results suggest insight into the biosynthetic capacities of the cultured marine actinomycetes. Thus, compounds **1–3**, **6**, and **7** represent interesting compounds for the design of novel and effective antibiotic drugs.

Supplementary Materials: The following are available online at http://www.mdpi.com/1660-3397/18/9/450/s1, Figure S1–S36: ^{1}H NMR, ^{13}NMR, ^{1}H-^{1}H COSY, multiplicity-edited HSQC, HMBC, and HRESIMS data of compounds 1-7.

Author Contributions: Conceived and designed the experiments, L.A.S. and D.T.A.Y.; prepared a large-scale culture and extracted the culture, S.S.E.; performed the experiments and purified the compounds, D.T.A.Y. and L.A.S.; interpreted the NMR and MS data, L.A.S., T.A.A. and D.T.A.Y.; wrote and revised the manuscript, L.A.S. and D.T.A.Y. All authors have read and agreed to the published version of the manuscript.

Funding: Deanship of Scientific Research (DSR) at King Abdulaziz University, Jeddah, under grant No. (G-270-141-39).

Acknowledgments: This project was funded by the Deanship of Scientific Research (DSR) at King Abdulaziz University, Jeddah, under grant No. (G-270-141-39). The authors, therefore, acknowledge with thanks the DSR for the technical and financial support. We would like to thank Rob van Soest for the identification of the sponge specimen.

Conflicts of Interest: The authors declare no conflict of interest.

References

1. Faulkner, D.J.; Harper, K.K.; Haygood, M.G.; Salomon, C.E.; Schmidt, E.W. Symbiotic bacteria in sponges: Sources of bioactive substances. In *Drugs from the Sea*; Fusetani, N., Ed.; Karger: Basel, Switzerland, 2000; pp. 107–119.
2. Tsukimoto, M.; Nagaoka, M.; Shishido, Y.; Fujimoto, J.; Nishisaka, F.; Matsumoto, S.; Haruari, E.; Imada, C.; Matsuzaki, T. Bacterial production of the tunicate-derived antitumor cyclic depsipeptide didemnin B. *J. Nat. Prod.* **2011**, *74*, 2329–2331. [CrossRef]
3. Brinkmann, C.M.; Marker, A.; Kurtböke, D.I. An Overview on Marine sponge-symbiotic bacteria as unexhausted sources for natural product discovery. *Diversity* **2017**, *9*, 40. [CrossRef]
4. Carroll, A.R.; Copp, B.R.; Davis, R.A.; Keyzers, R.A.; Prinsep, M.R. Marine natural products. *Nat. Prod. Rep.* **2020**, *37*, 175–223. [CrossRef]
5. Berdy, J. Bioactive microbial metabolites. *J. Antibiot. Tokyo* **2005**, *58*, 1–26. [CrossRef]
6. Newman, D.J.; Cragg, G.M.; Snader, K.M. Natural products as sources of new drugs over the period 1981–2002. *J. Nat. Prod.* **2003**, *66*, 1022–1037. [CrossRef]
7. Jiménez-Esquilín, A.E.; Roane, E.T.M. Antifungal activities of actinomycete strains associated with high-altitude sagebrush rhizosphere. *J. Ind. Microbiol. Biotechnol.* **2005**, *32*, 378–381. [CrossRef]
8. Olanrewaju, O.S.; Babalola, O.O. *Streptomyces*: Implications and interactions in plant growth promotion. *Appl. Microbiol. Biotechnol.* **2019**, *103*, 1179–1188. [CrossRef]
9. Ramesh, S.; Rajesh, M.; Mathivanan, N. Characterization of a thermostable alkaline protease produced by marine *Streptomyces fungicidicus* MML1614. *Bioprocess. Biosyst. Eng.* **2009**, *32*, 791–800. [CrossRef]
10. Prabavathy, V.R.; Mathivanan, N.; Murugesan, K. Control of blast and sheath blight diseases of rice using antifungal metabolites produced by *Streptomyces* sp. PM5. *Biol. Control* **2006**, *39*, 313–319. [CrossRef]
11. Prapagdee, B.; Kuekulvong, C.; Mongkolsuk, S. Antifungal potential of extracellular metabolites produced by *Streptomyces hygroscopicus* against phytopathogenic fungi. *Int. J. Biol. Sci.* **2008**, *4*, 330–337. [CrossRef]
12. Lam, K.S. Discovery of novel metabolites from marine actinomycetes. *Curr. Opin. Microbiol.* **2006**, *9*, 245–251. [CrossRef] [PubMed]
13. Hong, K.; Gao, A.H.; Xie, Q.Y.; Gao, H.; Zhuang, L.; Lin, H.P.; Yu, H.P.; Li, J.; Yao, X.S.; Goodfellow, M.; et al. Actinomycetes for marine drug discovery isolated from mangrove soils and plants in China. *Mar. Drugs* **2009**, *7*, 24–44. [CrossRef] [PubMed]
14. Pimentel-Elardo, S.M.; Kozytska, S.; Bugni, T.S.; Ireland, C.M.; Moll, H.; Hentschel, U. Antiparasitic compounds from *Streptomyces* sp. strains isolated from Mediterranean sponges. *Mar. Drugs* **2010**, *8*, 373–380. [CrossRef] [PubMed]
15. Renner, M.K.; Shen, Y.C.; Cheng, X.C.; Jensen, P.R.; Frankmoelle, W.; Kauffman, C.A.; Fenical, W.; Lobkovsky, E.; Clardy, J. Cyclomarins A-C, new anti-inflammatory cyclic peptides produced by a marine bacterium (*Streptomyces* sp.). *J. Am. Chem. Soc.* **1999**, *121*, 11273–11276. [CrossRef]
16. Xu, Y.; He, H.; Schulz, S.; Liu, X.; Fusetani, N.; Xiong, H.; Xiao, X.; Qian, P.Y. Potent antifouling compounds produced by marine *Streptomyces*. *Bioresour. Technol.* **2010**, *101*, 1331–1336. [CrossRef]
17. Sacramento, D.R.; Coelho, R.R.R.; Wigg, M.D.; Linhares, L.F.T.L.; Santos, M.G.M.; Semedo, L.T.A.S.; da Silva, A.J.R. Antimicrobial and antiviral activities of an actinomycete (*Streptomyces* sp.) isolated from a Brazilian tropical forest soil. *World J. Microbiol. Biotechnol.* **2004**, *20*, 225–229. [CrossRef]
18. Rahman, H.; Austin, B.; Mitchell, W.J.; Morris, P.C.; Jamieson, D.J.; Adams, D.R.; Spragg, A.M.; Schweizer, M. Novel anti-infective compounds from marine bacteria. *Mar. Drugs* **2010**, *8*, 498–518. [CrossRef]
19. Williams, S.T.; Goodfellow, M.; Alderson, G.; Wellington, E.M.H.; Sneath, P.H.A.; Sackin, M.J. A probability matrix for identification of some *Streptomyces*. *J. Gen. Microbiol.* **1983**, *129*, 1815–1830.

20. Tamehiro, N.; Hosaka, T.; Xu, J.; Hu, H.; Otake, N.; Ochi, K. Innovative approach for improvement of an antibiotic–overproducing industrial strain of *Streptomyces albus*. *Appl. Environ. Microbiol.* **2003**, *69*, 6412–6417. [CrossRef]

21. Higginbotham, S.J.; Murphy, C.D. Identification and characterisation of a *Streptomyces* sp. isolate exhibiting activity against methicillin-resistant *Staphylococcus aureus*. *Microbiol. Res.* **2010**, *165*, 82–86. [CrossRef]

22. Sousa, C.S.; Soares, A.C.F.; Garrido, M.S. Characterization of streptomycetes with potential to promote plant growth and biocontrol. *Sci. Agric. Piracicaba Braz.* **2008**, *65*, 50–55. [CrossRef]

23. Atta, H.M. Production of vitamin B12 by *Streptomyces fulvissimus*. *Egypt. J. Biomed. Sci.* **2007**, *23*, 166–184.

24. Cal, S.; Aparicio, J.F.; De Los Reyes-Gavilan, C.G.; Nicieza, J.; Sanchez, A. A novel exocytoplasmic endonuclease from *Streptomyces antibioticus*. *J. Biochem.* **1995**, *306*, 93–100. [CrossRef]

25. Brunakova, Z.; Godnay, A.; Timko, J. An extracellular endodeoxyribonuclease from *Sterptomyces aureofaciens*. *Biochim. Biophys. Acta* **2004**, *1721*, 116–123. [CrossRef]

26. Ramesh, S.; Mathivanan, N. Screening of marine actinomycetes isolated from the Bay of Bengal, India for antimicrobial activity and industrial enzymes. *World J. Microbiol. Biotechnol.* **2009**, *25*, 2103–2111. [CrossRef]

27. Nicieza, R.G.; Huergo, J.; Connolly, B.A.; Sanchex, J. Purification, characterization and role of nucleases and serine proteases in *Streptomyces* differentiation. *J. Biol. Chem.* **1999**, *274*, 20366–20375. [CrossRef]

28. Magarvey, N.A.; Keller, J.M.; Bernan, V.; Dworkin, M.; Sherman, D.H. Isolation and characterization of novel marine-derived actinomycete taxa rich in bioactive metabolites. *Appl. Environ. Microbiol.* **2004**, *70*, 7520–7529. [CrossRef]

29. Fenical, W.; Jensen, P.R. Developing a new resource for drug discovery: Marine actinomycete bacteria. *Nat. Chem. Biol.* **2006**, *2*, 666–673. [CrossRef]

30. Jensen, P.R.; Williams, P.G.; Oh, D.C.; Zeigler, L.; Fenical, W. Species-specific secondary metabolite production in marine actinomycetes of the genus *Salinispora*. *Appl. Environ. Microbiol.* **2007**, *73*, 1146–1152. [CrossRef]

31. Zazopoulos, E.; Huang, K.; Staffa, A.; Liu, W.; Bachmann, B.O.; Nonaka, K.; Ahlert, J.; Thorson, J.S.; Shen, B.; Farnet, C.M. A genomics-guided approach for discovering and expressing cryptic metabolic pathways. *Nat. Biotechnol.* **2003**, *21*, 187–190. [CrossRef]

32. McAlpine, J.B.; Bachmann, B.O.; Piraee, M.; Tremblay, S.; Alarco, A.M.; Zazopoulos, E.; Farnet, C.M. Microbial genomics as a guide to drug discovery and structural elucidation: ECO-02301, a novel antifungal agent, as an example. *J. Nat. Prod.* **2005**, *68*, 493–496. [CrossRef] [PubMed]

33. Udwary, D.W.; Zeigler, L.; Asolkar, R.N.; Singan, V.; Lapidus, A.; Fenical, W.; Jensen, P.R.; Moore, B.S. Genome sequencing reveals complex secondary metabolome in the marine actinomycete *Salinispora tropica*. *Proc. Natl. Acad. Sci. USA* **2007**, *104*, 10376–10381. [CrossRef] [PubMed]

34. Sakata, T.; Winzeler, E.A. Genomics, systems biology and drug development for infectious diseases. *Mol. Biosyst.* **2007**, *3*, 841–848. [CrossRef] [PubMed]

35. Shaala, L.A.; Youssef, D.T.A.; Badr, J.M.; Harakeh, S.M. Bioactive 2(1H)-pyrazinones and diketopiperazine alkaloids from a tunicate-derived actinomycete *Streptomyces* sp. *Molecules* **2016**, *21*, 1116. [CrossRef] [PubMed]

36. Shaala, L.A.; Youssef, D.T.A.; Badr, J.M.; Harakeh, S.M.; Genta-Jouve, G. Bioactive diketopiperazines and nucleoside derivatives from a sponge-derived *Streptomyces* species. *Mar. Drugs* **2019**, *17*, 584. [CrossRef] [PubMed]

37. Narayana, K.J.P.; Prabhakar, P.; Vijayalakshm, M.; Venkateswarlu, Y.; Krishna, P.S.J. Biological activity of phenylpropionic acid isolated from a terrestrial *Streptomycetes*. *Pol. J. Microbiol.* **2007**, *56*, 191–197. [PubMed]

38. Adamczeski, M.; Quinoa, E.; Crews, P. Novel sponge-derived amino acids. 5. Structures, stereochemistry, and synthesis of several new heterocycles. *J. Am. Chem. Soc.* **1989**, *111*, 647–654. [CrossRef]

39. Li, X.-Y.; Wang, Y.-H.; Yang, J.; Cui, W.-Y.; He, P.-J.; Munir, S.; He, P.-F.; Wu, Y.-X.; He, Y.-Q. Acaricidal activity of cyclodipeptides from *Bacillus amyloliquefaciens* W1 against *Tetranychus urticae*. *J. Agric. Food Chem.* **2018**, *66*, 10163–10168. [CrossRef]

40. Shigemori, H.; Tenma, M.; Shimazaki, K.; Kobayashi, J. Three new metabolites from the marine yeast *Aureobasidium pullulans*. *J. Nat. Prod.* **1998**, *61*, 696–698. [CrossRef]

41. De Rosa, S.; Mitova, M.; Tommonaro, G. Marine bacteria associated with sponge as source of cyclic peptides. *Biomol. Eng.* **2003**, *20*, 311–316. [CrossRef]

42. Restrepo, M.P.; Jaramillo, E.G.; Martínez, A.M.; Arango, A.M.; Restrepo, S.R. Anti-parasite and cytotoxic activities of chloro and bromo L-tyrosine derivatives. *J. Braz. Chem. Soc.* **2018**, *29*, 2569–2579. [CrossRef]

43. Barros, J.; Serrani-Yarce, J.C.; Chen, F.; Baxter, D.; Venables, B.J.; Dixon, R.A. Role of bifunctional ammonia-lyase in grass cell wall biosynthesis. *Nat. Plants* **2016**, *2*, 16050. [CrossRef] [PubMed]
44. Vogt, T. Phenylpropanoid biosynthesis. *Mol. Plant* **2010**, *3*, 2–20. [CrossRef] [PubMed]
45. Jain, S.; Laphookhieo, S.; Shi, Z.; Fu, L.; Akiyama, S.; Chen, Z.; Youssef, D.T.A.; van Soest, R.W.M.; El Sayed, K.A. Reversal of P-glycoprotein-mediated multidrug resistance by sipholane triterpenoids. *J. Nat. Prod.* **2007**, *70*, 928–931. [CrossRef]
46. Youssef, D.T.A.; Ibrahim, A.K.; Khalifa, S.I.; Mesbah, M.K.; Mayer, A.M.S.; van Soest, R.W.M. New anti-inflammatory sterols from the Red Sea sponges *Scalarispongia aqabaensis* and *Callyspongia siphonella*. *Nat. Prod. Commun.* **2010**, *5*, 27–31.
47. Küster, E. Outline of a comparative study of criteria used in characterization of the actinomycetes. *Int. Bull. Bacteriol. Nomencl. Taxon.* **1959**, *9*, 97–104. [CrossRef]
48. Acar, J.F. The disc susceptibility test. In *Antibiotics in Laboratory Medicine*; Lorian, V., Ed.; Williams and Wilkins: Baltimore, MD, USA, 1980; pp. 24–54.
49. Kiehlbauch, J.A.; Hannett, G.E.; Salfinger, M.; Archinal, W.; Monserrat, C.; Carlyn, C. Use of the National Committee for Clinical Laboratory Standards Guidelines for Disk Diffusion Susceptibility Testing in New York State Laboratories. *J. Clin. Microbiol.* **2000**, *38*, 3341–3348. [CrossRef]
50. Shaala, L.A.; Khalifa, S.I.; Mesbah, M.K.; van Soest, R.W.M.; Youssef, D.T.A. Subereaphenol A, a new cytotoxic and antimicrobial dibrominated phenol from the Red Sea sponge *Suberea mollis*. *Nat. Prod. Commun.* **2008**, *3*, 219–222. [CrossRef]
51. CLSI, Clinical and Laboratory Standards Institute. *Performance Standards for Antimicrobial Disk Susceptibility Tests*, 9th ed.; CLSI Documents M07-A9; West Valley Road: Wayne, PA, USA, 2007.

 *marine drugs*

Article

Thalassosterol, a New Cytotoxic Aromatase Inhibitor Ergosterol Derivative from the Red Sea Seagrass *Thalassodendron ciliatum*

Reda F. A. Abdelhameed [1,†], Eman S. Habib [1,†], Marwa S. Goda [1], John Refaat Fahim [2], Hashem A. Hassanean [1], Enas E. Eltamany [1], Amany K. Ibrahim [1], Asmaa M. AboulMagd [3], Shaimaa Fayez [4,5], Adel M. Abd El-kader [6,7], Tarfah Al-Warhi [8], Gerhard Bringmann [4,*], Safwat A. Ahmed [1,*] and Usama Ramadan Abdelmohsen [2,6]

[1] Department of Pharmacognosy, Faculty of Pharmacy, Suez Canal University, Ismailia 41522, Egypt; omarreda_70@yahoo.com (R.F.A.A.); emansnd@yahoo.com (E.S.H.); marwa_saeed@pharm.suez.edu.eg (M.S.G.); Drhashemhassanean2020@pharm.suez.edu.eg (H.A.H.); enastamany@gmail.com (E.E.E.); am_kamal66@yahoo.com (A.K.I.)

[2] Department of Pharmacognosy, Faculty of Pharmacy, Minia University, Minia 61519, Egypt; john.mcconnell@lancet.com (J.R.F.); usama.ramadan@mu.edu.eg (U.R.A.)

[3] Pharmaceutical Chemistry Department, Faculty of Pharmacy, Nahda University, BeniSuef 62513, Egypt; asmaa_owis@yahoo.com

[4] Institute of Organic Chemistry, University of Würzburg, Am Hubland, 97074 Würzburg, Germany; shaimaa.seaf@uni-wuerzburg.de

[5] Department of Pharmacognosy, Faculty of Pharmacy, Ain-Shams University, Cairo 11566, Egypt

[6] Department of Pharmacognosy, Faculty of Pharmacy, Deraya University, New Minia 61111, Egypt; ad_cognosy@yahoo.com

[7] Department of Pharmacognosy, Faculty of Pharmacy, Al-Azhar University, Assiut 71524, Egypt

[8] Department of Chemistry, College of Science, Princess Nourah bint Abdulrahman University, Riyadh 13414, Saudi Arabia; tarfah-w@hotmail.com

* Correspondence: bringmann@chemie.uni-wuerzburg.de (G.B.); safwat_aa@yahoo.com (S.A.A.); Tel.: +49-0931-3185323 (G.B.); +20-010-92638387 (S.A.A.)

† Equal contributions.

Received: 18 June 2020; Accepted: 7 July 2020; Published: 8 July 2020

Abstract: *Thalassodendron ciliatum* (Forssk.) Den Hartog is a seagrass belonging to the plant family Cymodoceaceae with ubiquitous phytoconstituents and important pharmacological potential, including antioxidant, antiviral, and cytotoxic activities. In this work, a new ergosterol derivative named thalassosterol (**1**) was isolated from the methanolic extract of *T. ciliatum* growing in the Red Sea, along with two known first-reported sterols, namely ergosterol (**2**) and stigmasterol (**3**), using different chromatographic techniques. The structure of the new compound was established based on 1D and 2D NMR spectroscopy and high-resolution mass spectrometry (HR-MS) and by comparison with the literature data. The new ergosterol derivative showed significant in vitro antiproliferative potential against the human cervical cancer cell line (HeLa) and human breast cancer (MCF-7) cell lines, with IC_{50} values of 8.12 and 14.24 µM, respectively. In addition, docking studies on the new sterol **1** explained the possible binding interactions with an aromatase enzyme; this inhibition is beneficial in both cervical and breast cancer therapy. A metabolic analysis of the crude extract of *T. ciliatum* using liquid chromatography combined with high-resolution electrospray ionization mass spectrometry (LC-ESI-HR-MS) revealed the presence of an array of phenolic compounds, sterols and ceramides, as well as di- and triglycerides.

Keywords: cytotoxic activity; ergosterol derivative; metabolic analysis; docking studies; seagrass; *Thalassodendron ciliatum*

1. Introduction

Seagrasses are marine flowering plants that grow underwater along temperate and tropical coastlines, providing shelter or food for other marine organisms. They are important components of the near-shore ecosystem and are used as biological indicators of environmental quality [1]. In East Africa, seagrasses are used as fertilizers and medicinally for the management of fever and skin diseases [2]. *Thalassodendron ciliatum* (Forssk.) Den Hartog (Family Cymodoceaceae) is a sub-tidal or shallow-depositional seagrass species that is generally found in extensive and monotonous meadows. This sickle-leaved cymodocea, commonly known as "Majani kumbi", grows in the Red Sea, the western Indian Ocean, and the Indo-Pacific region [3]. *T. ciliatum* has both horizontal and vertical rhizomes, with a cluster of leaves at the top of each living stem [4]. The leaves of *T. ciliatum* are characterized by the presence of many tannin-containing cells and are rich in phenolic constituents [5]. Pharmacological studies showed that *T. ciliatum* possesses antioxidant, antiviral, and cytotoxic activities, which are highly correlated to its flavonoid content [6–8]. Several flavonoids were reported from *T. ciliatum* such as quercetin 3-*O*-β-D-xylopyranoside, asebotin, 3-hydroxyasebotin, rutin, and racemic catechin [6]. Recently, a new diglyceride ester and asebotin were isolated from *T. ciliatum*, showing antiviral activities against the H5N1 virus through inhibition of the virus titre by 67.26% and 53.81%, respectively [7]. A new dihydrochalcone diglycoside was also identified from *T. ciliatum*, exhibiting anti-influenza A virus activity [9]. In 2018, our group isolated a cytotoxic phytosphingosine-type ceramide from *T. ciliatum* of the Red Sea in addition to different sterols with anti-inflammatory effects [10]. Gas chromatography coupled to mass spectrometry analysis (GC-MS coupling) was performed for identification of the lipoidal matter of the *n*-hexane fraction of *T. ciliatum*, revealing the presence of saturated and unsaturated long-chain fatty acids, such as tetradecanoic acid, eicosanoic acid, 9,12-hexadecadienoic acid, and 8,11,14- eicosatrienoic acid, in addition to other volatile compounds, as well as 1-heneicosanol, 2,6-bis (1,1-dimethylethyl)phenol, and 1-tridecanol [11]. Some genera of Cymodoceaceae showed different steroidal profiles, among them compounds like cholesterol, β-sitosterol, stigmasterol, campesterol, and 22,23-dihydrobrassicasterol. Four 3-ketosteroids with a 24-ethyl cholestane side chain were also isolated from *Cymodocea nodosa* [12,13]. A comparative study on five seagrasses of different families revealed a high percentage of stigmasterol and β-sitosterol in both *Cymodocea serrulate* and *Halodule uninervis* [14]. Our previous study was the only one that reported on the isolation of different steroids from *T. ciliatum* with anti-inflammatory effects, including 7β-hydroxy cholesterol, 7β-hydroxysitosterol, stigmasterol glucoside, and β-sitosterol glucoside [10]. Therefore, we herein continue our chemical and biological investigation on *T. ciliatum* growing in the Red Sea, along with high-resolution electrospray ionization mass spectrometry (LC-ESI-HR-MS)-assisted metabolic profiling of this important seagrass.

2. Results and Discussion

2.1. Metabolic Profiling

Metabolic profiling of the crude extract of *T. ciliatum* using the LC-ESI-HR-MS technique (Figures S1 and S2) revealed the presence of a broad variety of metabolites such as flavonoids, chalcones, phenolic acids, sterols, fatty acids, anthraquinones, and terpenoids. The LC-ESI-HR-MS profiling for the rapid identification of natural metabolites resulted in the annotation of 20 compounds identified by comparison of their data, particularly their accurate masses, with those from some databases, e.g., the Dictionary of Natural Products (DNP) and the Metabolite and Chemical Entity (METLIN) database, as shown in Figure 1. Mass accuracy was calculated as [measured mass-expected mass/expected mass] $\times 10^6$ and expressed in parts per million (ppm) error [15]. The herein-characterized metabolites of *T. ciliatum* were found to be in accordance with those obtained in previous phytochemical studies, which reported the isolation of several phenolic compounds and sterols [6,10]. From Table 1, protocatechuic acid, butein, kaempferol, catechin, 1,4,5-trihydroxy-7-methoxy-3-methyl anthraquinone, quercetin, 2-ω-dihydroxy emodin, isorhamnetin, quercetin-3-*O*-β-D-xylopyranoside, asebotin, and rutin were reported to have antioxidant activity that can protect from cardiovascular diseases, liver damage, and proliferation of abnormal cells [6,16–20]. Butein showed an aromatase inhibition activity that can be effective in breast cancer treatment [21]. A long-chain polyunsaturated fatty acid, namely 6*E*,8*E*,10*E*,12*E*-octadecatetraenoic acid, as well as the diterpene linearol, exhibited antimicrobial and anticancer activities [22,23]. Likewise, both sphaerollane I and ergosterol inhibited proliferation of various malignant cell lines [24,25]. Linearol and rutin were reported to have antioxidant, cytotoxic and antiviral activities [18,23]. It is worth noting that the aforementioned antioxidant, antiviral, and cytotoxic activities of the crude extract of *T. ciliatum* may be correlated to the activities of the identified metabolites.

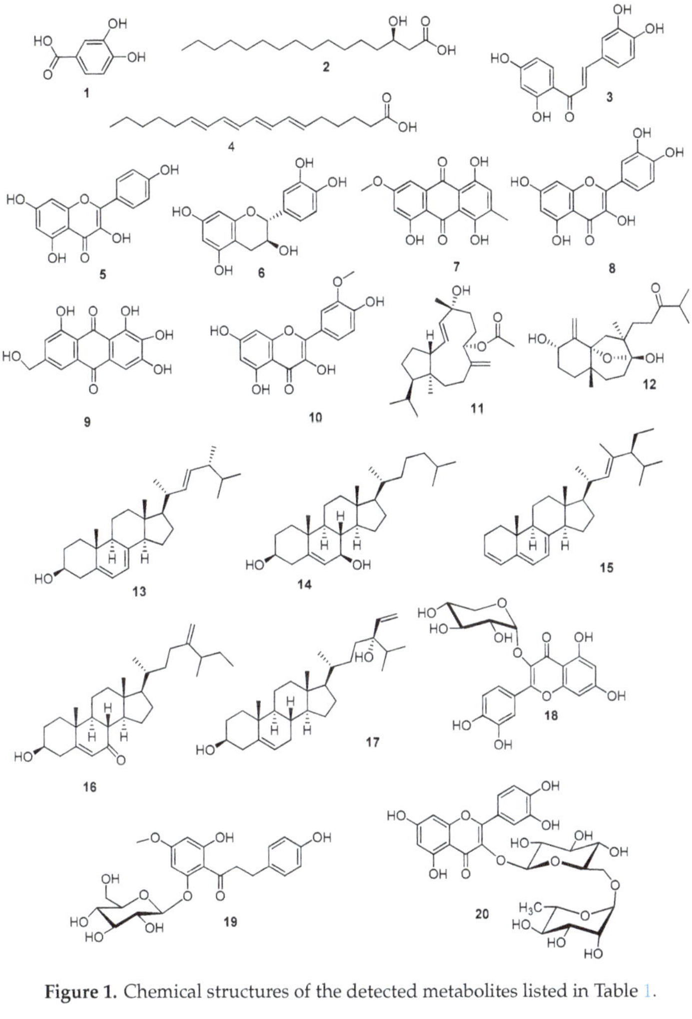

Figure 1. Chemical structures of the detected metabolites listed in Table 1.

Table 1. Metabolic analysis of a crude extract of *Thallasodendron ciliatum*.

	Polarity Mode	Retention Time (min)	MZmine ID	m/z *	Measured Mass	Expected Mass	Mass Error (ppm)	Name	Molecular Formula **	Source	Ref.
1	positive	1.51	9930	155.0347	154.0274	154.0266	5.2	Protocatechuic acid	$C_7H_6O_4$	*Allium cepa*	[26]
2	negative	11.29	382	271.2269	272.2342	272.2351	−3.3	3R-Hydroxypalmitic acid	$C_{16}H_{32}O_3$	*Saccharomycopsis* sp.	[27]
3	positive	5.07	9019	273.0748	272.0675	272.0685	−3.7	Butein	$C_{15}H_{12}O_5$	*Dalbergia odorifera*	[17]
4	positive	5.63	9699	277.2153	276.2080	276.2089	−3.3	6E,8E,10E,12E-Octadecatetraenoic acid	$C_{18}H_{28}O_2$	*Anadyomene stellata*	[28]
5	negative	4.46	3937	285.0393	286.0466	286.0477	−3.8	Kaempferol	$C_{15}H_{10}O_6$	*Fragaria chiloensis*	[29]
6	negative	2.18	3954	289.0705	290.0778	290.0790	−4.1	Catechin	$C_{15}H_{14}O_6$	*T. ciliatum*	[6]
7	negative	5.43	4767	299.0551	300.0623	300.0634	−3.7	1,4,5-Trihydroxy-7-methoxy-3-methylanthraquinone	$C_{16}H_{12}O_6$	*Chaetomium globosum*	[30]
8	negative	5.07	4942	301.0341	302.0414	302.0427	−4.3	Quercetin	$C_{15}H_{10}O_7$	*Fragaria chiloensis*	[29]
9	positive	3.37	9298	303.0503	302.0430	302.0427	0.9	2-ω-Dihydroxyemodin	$C_{15}H_{10}O_7$	*Aspergillus nidulans*	[31]
10	negative	4.76	4129	315.0497	316.0570	316.0583	−4.1	Isorhamnetin	$C_{16}H_{12}O_7$	*Stigma maydis*	[32]
11	negative	13.90	4382	347.2581	348.2654	348.2664	−2.9	Sphaerollane I	$C_{22}H_{36}O_3$	*Sphaerocos coronopifolis*	[33]
12	positive	8.66	12289	353.2706	352.2634	352.2614	5.7	Linearol	$C_{21}H_{36}O_4$	*Sideritis condensata*	[23]
13	positive	12.71	9842	397.3452	396.3379	396.3392	−3.3	Ergosterol (**2**)	$C_{28}H_{44}O$	*Ganoderma lucidum*	[25]
14	positive	13.60	10050	403.3528	402.3456	402.3498	−10.4	7β-Hydroxycholesterol	$C_{27}H_{46}O_2$	*T. ciliatum*	[10]
15	positive	12.49	8715	407.3677	406.3604	406.3600	0.9	23-Methylstigmasta-3Z,5Z,7Z,22E-tetraene	$C_{30}H_{46}$	*Suillus luteus*	[34]
16	positive	10.46	8941	427.3572	426.3499	426.3498	0.2	26-Methylergosta-5,24(28)-diene-7-one-3-ol	$C_{29}H_{46}O_2$	*Geodia japonica*	[35]
17	positive	15.43	320	429.3732	428.3659	428.3654	1.2	24R-Stigmasta-5,28-diene-3β,24 -diol	$C_{29}H_{48}O_2$	*Sargassum fusiforme*	[36]
18	negative	3.37	4110	433.0765	434.0838	434.0849	−2.5	Quercetin-3-O-β-D-xylopyranoside	$C_{20}H_{18}O_{11}$	*T. ciliatum*	[6]
19	positive	6.10	705	451.1642	450.1570	450.1526	9.8	Asebotin	$C_{22}H_{26}O_{10}$	*T. ciliatum*	[6]
20	negative	2.33	5311	609.1466	610.1539	610.1534	0.8	Rutin	$C_{27}H_{30}O_{16}$	*T. ciliatum*	[6]

* m/z is expressed in negative or positive formula; ** Molecular formula is expressed in a neutral formula.

2.2. Identification of the Isolated Metabolites

Compound **1** (Figure 2) was obtained as a white powder and its molecular formula was determined by ESI-HR-MS as $C_{30}H_{49}O_5$ (found at *m/z* 489.3573 [M − H]⁻ (Figure S3), calculated for 489.3579), indicating the presence of six degrees of unsaturation in the molecule. The NMR spectral data of compound **1** (Table 2) (Figures S4–S7) also closely matched those reported in the literature for ergostane-type steroids [27,37], showing signals characteristic of four secondary methyl groups resonating at δ_H 1.06 (CH_3-21), 0.92 (CH_3-28), 0.877 (CH_3-27), and 0.873 (CH_3-26), as well as one tertiary methyl group at δ_H 1.03 (CH_3-19), along with their carbon resonances at δ_C 19.9, 18.4, 22.9, 23.1, and 14.1, respectively. The two geminally coupled (*J* = 12.2 Hz) one-proton doublets at δ_H 3.75 and 3.65, and the corresponding oxygenated methylene carbon at δ_C 60.6 were indicative of the presence of a hydroxy group at C-18, which was corroborated by the observed Heteronuclear Multiple Bond Correlation (HMBC) correlations of H₂-18 with C-12, C-13, and C-17 (Figure 3) (Figure S8). Additionally, the NMR data of compound **1** displayed two one-proton doublets of doublets at δ_H 5.12 (*J* = 15.2, 6.5 Hz) and 5.23 (*J* = 15.2, 7.2 Hz), ascribable to the olefinic protons H-22 and H-23, together with their corresponding methine carbons at δ_C 137.5 and 133.1, respectively.

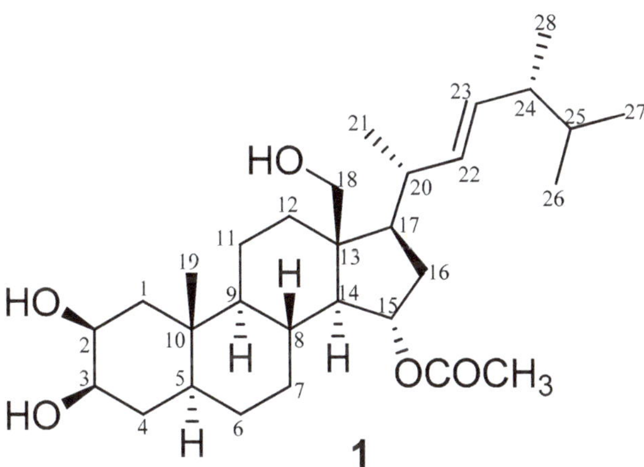

Figure 2. Chemical structure of the new ergosterol derivative **1** (2β,18-dihydroxy-15α-acetoxy-5,6,7,8-tetrahydroergosterol), named thalassosterol.

Table 2. ¹H (400 MHz) and ¹³C (100 MHz) NMR spectroscopic data of compound **1** (CD₃OD, δ in ppm, *J* in Hz).

No.	δ_H	δ_C
1	1.42, 2.09, m	39.1
2	4.72, br. d (*J* = 2.2)	76.0
3	4.75, br. d (*J* = 2.2)	76.3
4	1.60, 1.80, m	30.4
5	1.59, m	40.2
6	1.25, 1.50, m	28.9
7	0.96, 1.59, m	32.3
8	1.74, m	32.4
9	0.83, m	56.8
10	—	36.5

Table 2. *Cont.*

No.	δ_H	δ_C
11	1.37, 1.56, m	21.8
12	0.96, 2.62, m	36.2
13	—	48.8
14	1.20, m	59.8
15	4.99, m	74.4
16	1.30, 2.46, m	39.9
17	1.14, m	58.1
18	3.65, d (J = 12.2) 3.75 d (J = 12.2)	60.6
19	1.03, br. s	14.2
20	1.72, m	36.6
21	1.06, d (J = 6.4)	19.9
22	5.15, dd (J = 15.2, 6.5)	137.5
23	5.21, dd (J = 15.2, 7.2)	133.2
24	1.84, m	44.4
25	1.51, m	29.2
26	0.873, d (J = 6.64)	23.2
27	0.877, d (J = 6.64)	23.0
28	0.92, d (J = 6.8)	18.4
-CO-CH$_3$	2.00, s	21.2
-CO-CH$_3$	—	172.6

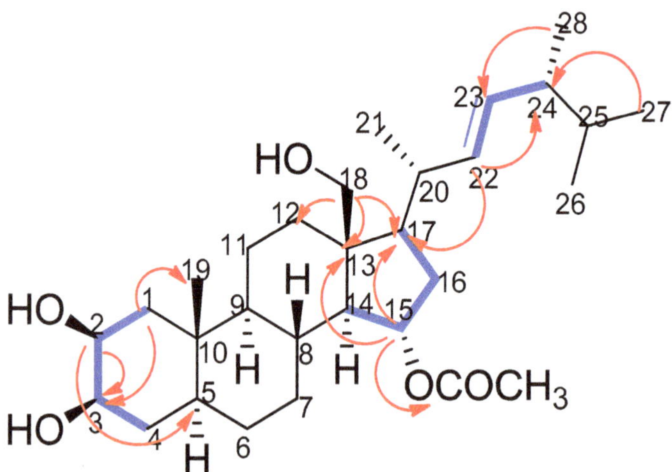

Figure 3. Key ^{1}H-^{1}H COSY (●—) and Heteronuclear Multiple Bond Correlation (HMBC) (⌒→) interactions in thalassosterol (**1**), (2β,18-dihydroxy-15α-acetoxy-5,6,7,8-tetrahydroergosterol). Blue bold line represents ^{1}H-^{1}H COSY correlations between H1/H2, H2/H3, H3/H4, H14/H15, H15/H16, H16/H17, H22/H23 and H23/H24, while red single-head arrow represents HMBC correlations between H1/C19, H1/C3, H2/C3, H2/C5, H15/C13, H15/C17, H15/CO, H18/C12, H18/C17, H18/C13, H22/C17, H22/C24, H27/C24 and H28/C23.

The position of this double bond between C-22 and C-23 was also supported by the obtained Heteronuclear Multiple Bond Correlation (HMBC) cross peaks of H-22/C-17 and H-22/C-24, as well as by the proton/proton Correlation Spectroscopy (^{1}H-^{1}H COSY) of H-22/H-23 and H-23/H-24 (Figure 3 and Figure S9), whereas the large coupling constant (15.2 Hz) between H-22 and H-23 typically indicated a trans-configuration of that double bond. Moreover, the ^{1}H NMR signals at δ_H 4.71 (1H, br. d, J = 2.2 Hz) and 4.74 (1H, br. d, J = 2.2 Hz) were attributed to H-2 and H-3, respectively, suggesting that the hydroxylation in ring A of this sterol should be at C-2 and C-3, which was also confirmed by Heteronuclear Single Quantum Correlation (HSQC). These assignments were also substantiated with the aid of ^{1}H-^{1}H COSY, HSQC, and HMBC analyses (Figure 3).

The β-configuration of the hydroxy groups at C-2 and C-3 was clearly deduced from the observed chemical shifts of both carbons (δ_C 75.9 and 76.3, respectively) [38], as well as through the Nuclear Overhauser Effect Spectroscopy (NOESY) correlations observed between H-2/H-3, H-2/H-5, H-2/H-4a, H-3/H-1a, H-3/H-2, and H-3/H-5 (Figure 4 and Figure S10). On the other hand, the three-proton singlet at δ_H 2.0, along with its corresponding carbon resonating at δ_C 21.2 and the quaternary carbonyl signal at δ_C 172.6 were consistent with the presence of an acetoxy group, which was unambiguously allocated at C-15 based on the marked downfield shift of both H-15 (δ_H 4.98) and C-15 (δ_C 74.3) in comparison with ergosterol (**2**) and other related ergostane derivatives [25,37]. The noticed HMBC correlation of H-15 with the carbonyl carbon of this acetoxy group, in addition to its three-bond connectivities to both C-13 and C-17, were also in good agreement with the acetylation of compound **1** at C-15, while the observed NOESY correlations between H-15 and both H-7b and H-8 verified the α-orientation of that acetoxy group (Figure 4). Based on the above-mentioned assignments, which were totally supported by the DEPT-135, ^{1}H-^{1}H COSY, HSQC, HMBC, and NOESY experiments, compound **1** was identified as 2β,18-dihydroxy-15α-acetoxy-5,6,7,8-tetrahydroergosterol, or, in other words, as (2β,3β-dihydroxy-13β-hydroxymethyl-10β-methyl-17β-(1,4,5-trimethyl-hex-2E-enyl)-hexadecahydro-cyclopenta[α]phenanthren-15α-yl acetate) (Figure 2). To the best of our knowledge, this molecule is a new compound, henceforth named thalassosterol.

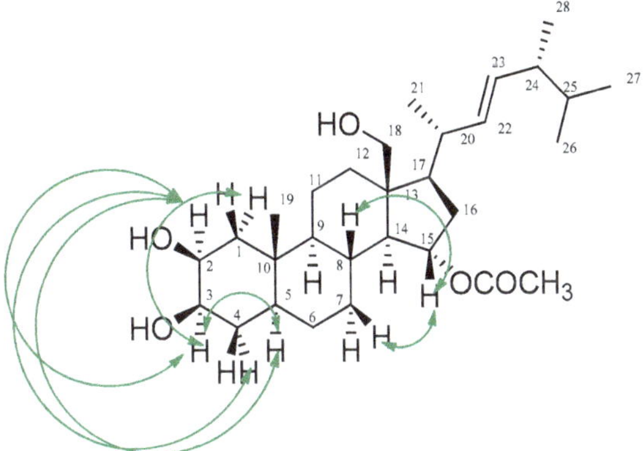

Figure 4. Key Nuclear Overhauser Effect Spectroscopy (NOESY) ( ) correlations of thalassosterol (**1**), (2β,18-dihydroxy-15α-acetoxy-5,6,7,8-tetrahydroergosterol). Green double-head arrow represents NOESY correlations between H2/H3, H2/H5, H2/H4a, H3/H1a, H3/H2, H3/H5, H15/H7b and H15/H8.

Compounds **2** and **3** (Figure 5) were identified as the known steroids ergosterol (**2**) and stigmasterol (**3**), respectively, by comparing their spectral data with those reported in the literature [25,37,39,40]. Both compounds were also detected in the metabolic analysis of *T. ciliatum*. It is worth mentioning that it is the first time to report on the isolation of both ergosterol and stigmasterol from the seagrass *T. ciliatum*.

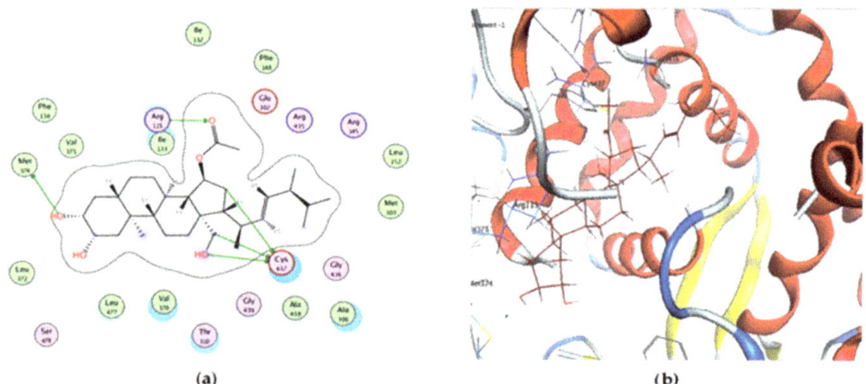

Figure 5. Chemical structures of the known metabolites ergosterol (**2**) and stigmasterol (**3**) isolated from *T. ciliatum*.

2.3. Docking Study

It can be observed generally through looking at the best score poses that binding of the new ergosterol, thalassosterol (**1**), at the active site is similar to that of the endogenous aromatase substrate, exemestane [41]. The binding interaction shows that the steroidal nucleus, overlapped with the hydrophobic environment of the binding pocket with ring D, is oriented towards the Met 374 residue and the β-face positioned towards the heme moiety. It can be clearly noticed that, while the hydroxy group of the ring A engages one hydrogen bond donor with the backbone amide of Met 374 as reported (3.82 Å) [42], the C15-keto oxygen atom of ring D acts as a hydrogen bond acceptor from Arg 115 (3.12 Å). Moreover, three extra hydrogen bond donors were involved via the interaction with Cys 417 residue with binding interaction score −8.219 kcal/mol (Figure 6). It is worth to mention that the H-bond interaction of the targeted new compound with Met 374 represents the selectivity of the latter with the active site of the aromatase. Moreover, the binding interactions of the target ergosterol derivative with different amino acids of the binding site of the aromatase did not only confirm the selectivity but also the possible binding interactions when compared with exemestane.

(a)   (b)

Figure 6. (**a**) The 2D caption of thalassosterol (**1**), (2β,18-dihydroxy-15α-acetoxy-5,6,7,8-tetrahydroergosterol), binding to the active site of aromatase. (**b**) Binding pattern of compound **1** colored by element into the receptor binding site showing five interactions (dotted lines).

2.4. In Vitro Antiproliferative Activity

The SRB assay is a rapid, sensitive, and inexpensive method for screening for antitumor activities of chemical agents against different malignant cell lines [43]. The antitumor effects of the isolated compounds were assessed by determination of the concentration required for 50% of growth inhibition (IC$_{50}$). The cytotoxic potentials of the new compound **1** and of doxorubicin as a positive control are shown in Table 3 against different malignant cell lines, including the human cervical cancer cell line (HeLa), a human breast cancer cell line (MCF-7), and a human liver cancer cell line (HepG2). The efficacy of compound **1** follows the order of HeLa > MCF-7 > HepG2 malignant cell lines. The IC_{50} of compound **1** is 8.13 ± 0.21 µM (mean $\pm$ SD) against the HeLa cell line and this result is closely related to the one obtained with doxorubicin as a positive control (Table 3). Moreover, compound **1** showed a weak antiproliferative activity on cells of the normal Vero cell line (> 90 µM).

Table 3. IC_{50} values (µM) of thalassosterol (**1**) and doxorubicin against different human cell lines, HeLa, human liver cancer cell line (HepG2), and human breast cancer cell line (MCF-7).

	Human Cancer Cell Lines		
	HeLa IC50 (µM)	HepG2 IC50 (µM)	MCF-7 IC50 (µM)
Thalassosterol (1)	8.13 ± 0.21 *	48.64 ± 0.22 *	14.26 ± 0.40 *
Doxorubicin	6.77 ± 0.07	9.02 ± 0.06	8.65 ± 0.03

Human cervical cancer cell line (HeLa), human liver cancer cell line (HepG2), human breast cancer cell line (MCF-7). * Significantly different from doxorubicin as the positive control. Each data point represents the mean $\pm$ SD of three independent experiments (significant differences at $p < 0.05$).

According to the US NCI (National Cancer Institute) plant screening program guidelines, a crude extract is considered to have in vitro antiproliferative activity if the IC_{50} value after incubation between 48 and 72 h is smaller than 20 µg/mL, and less than 4 µg/mL for pure compounds [44]. Based on these NCI guidelines, the new compound **1** exhibited a high cytotoxic potential against the HeLa cell line, with an IC_{50} value of 8.13 ± 0.21 µM (less than 8.17 µM, equivalent to 4 µg/mL of the US NCI guidelines). On the other hand, compound **1** showed a weak cytotoxic activity against the HepG2 cell line, with an IC_{50} value of 48.64 ± 0.22 µM. Moreover, it was found to have a moderate cytotoxic activity against the breast cancer cell line MCF-7, with an IC_{50} value of 14.26 ± 0.40 µM when compared with doxorubicin as a positive control (8.65 ± 0.03 µM). This cytotoxic activity against the mentioned types of estrogen-responsive cancers, cervical cancer and breast cancer [41] was confirmed by the previously discussed docking study by blocking the aromatase enzyme that synthesizes estrogens.

3. Materials and Methods

3.1. Plant Material

The seagrass *T. ciliatum* was collected from Sharm El Sheikh at the Egyptian Red Sea, then air-dried and stored at a low temperature (-24 °C) until further processing. The plant was identified by Dr. Tarek Temraz, Marine Science Department, Faculty of Science, Suez Canal University, Ismailia, Egypt. A voucher sample (no. SAA-41) was deposited in the herbarium section of Pharmacognosy Department, Faculty of Pharmacy, Suez Canal University, Ismailia, Egypt.

3.2. General Experimental Procedures

^{1}H and ^{13}C NMR spectra were obtained with a Bruker Avance III HD 400 spectrometer operating at 400 MHz for ^{1}H and 100 MHz for ^{13}C. Both ^{1}H and ^{13}C NMR chemical shifts are expressed in δ values in regard to the solvent peaks δ_H 3.3 and δ_C 49 ppm for CD_3OD, and coupling constants are given in Hertz (Hz). Thin layer chromatography (TLC) analysis was carried out on aluminum-backed plates pre-coated with silica gel F_{254} (20 × 20 cm; 200 µm; 60 Å (Merck™, Darmstadt, Germany), while silica gel 60/230–400 µm mesh size (Whatman™, Sanford, ME, USA) was used for column chromatography. Sephadex® LH-20 (Sigma Aldrich, Bremen, Germany), and reversed-phase octadecyl silica (ODS) gel (YMC, Kyoto, Japan) were also utilized.

3.3. Extraction and Isolation

The air-dried material of *T. ciliatum* (90 g) was ground and extracted with a 1:1 mixture of CH_2Cl_2/MeOH (2 L × 3) at room temperature. The combined extracts were concentrated under vacuum to afford a dark-green residue (Tc, 21 g), which was then chromatographed over an open silica gel column using *n*-hexane/ethyl acetate (EtOAc) as the eluent (95:5-0:100) and EtOAc/methanol (90:10-50:50), with gradient elution giving seven fractions, Tc-A~Tc-G. The second fraction eluted using 25% EtOAc in hexane, Tc-B, was concentrated to afford a green residue (5 g) and was purified on silica gel using hexane/EtOAc (95:5) giving nine subfractions, Tc-B-1~Tc-B-9. Among them, subfractions Tc-B-2~Tc-B-4, having similar TLC patterns, were combined (Tc-B-1′, 90 mg) and re-chromatographed on silica gel using hexane-EtOAc (95:5) giving four sub-subfractions, Tc-B-1′-1~Tc-B-1′-4. One of the resulting sub-subfractions, Tc-B-1′-2, afforded compound **3** (11 mg). Another sub-subfraction, Tc-B-1′-3 (30 mg), was also applied to a Sephadex LH-20 column and eluted with $CHCl_3$-MeOH (1:1) to afford compounds **1** (14 mg, white powder) and **2** (16 mg). Compound **1** was finally purified on an open ODS column using MeOH/H_2O (8:2).

3.4. Metabolic Profiling

The metabolic study was performed using LC-ESI-HR-MS for dereplication purposes according to the literature [45]. An aliquot of 10 μL of the methanolic extract of *T. ciliatum* (1 mg/mL) was injected to an Accela HPLC (Thermo Fisher Scientific, Bremen, Germany) using an ACE C18 column of 75 mm × 3 mm having 5 μm internal diameter (Hichrome Limited, Reading, UK) and equipped with an Accela UV-visible detector. This compartment was coupled to an Exactive (Orbitrap) mass spectrometer (Thermo Fisher Scientific, Bremen, Germany). The gradient elution was done using purified water (total organic carbon 20 ppb) and acetonitrile; each containing 0.1% formic acid with a flow rate of 300 μL/min at ambient temperature. The gradient elution technique was assessed by increasing the concentration of acetonitrile from 10% to 100% within 30 min, followed by an isocratic period of 5 min and then reducing the concentration of acetonitrile to 10% within 1 min. ESI-HR-MS analysis was performed in both positive and negative ionization modes, with a spray voltage of 4.5 kV and a capillary temperature of 320 °C. The ESI-MS mass range was set at *m/z* 100–2000 using in-source collision-induced dissociation (CID) mechanism and *m/z* 50–1000. The raw HR-MS data were imported and analyzed using MZmine 2.12. Excel macros were also employed to dereplicate each *m/z* ion peak with metabolites in the database, using the retention time and an *m/z* threshold of ± 5 ppm, to attain the tentative identification of the compounds. A chemotaxonomic filter was applied to the obtained hits in order to limit the number of identities per metabolite and to only include the relevant ones. The compounds were identified by a comparison of their data, particularly their accurate masses, with those from some databases, e.g. DNP and METLIN. The retention time, *m/z*, molecular weight, molecular formula, MZmine ID, name, and biological source were determined.

3.5. Modeling Study on the Binding Between the New Ergosterol Derivative 1 and the Aromatase Binding Site

Molecular Operating Environment (MOE) was adopted for docking calculations. The structures of the compounds were generated by ChemDraw Ultra 11.0. Molecular docking calculations were applied on human placental aromatase (PDB code 3S7S, http://www.rcsb.org/pdb/home/home.do). Since the aromatase enzyme is complexed with the steroid exemestane as an inhibitor, this model was chosen for steroidal scaffold inhibitors. Docking simulation was done on the test compounds with the following protocol. Structure arrangement process was used to revise the protein errors, and a reasonable protein structure was set up on default rules on MOE. Finally, the Gasteiger methodology was used to calculate the partial charges of the protein [46]. The ligands were protonated and the correction of atom and bond types were defined, hydrogen atoms were added, and finally minimization was performed (MMFF94x, gradient: 0.01).

The default Triangle Matcher placement method was selected for docking. The GBVI/WSA dG scoring function that determines the free energy of binding of the ligand from a given pose, was chosen to rank the final poses. The ligand complex with the enzyme having the lowest S-score was selected. The redocking of ligand with its target revealed an RMSD of 0.98 Å, which confirmed that the ligand binds to the same pocket and assured the dependability of parameters of docking.

3.6. *In Vitro Antiproliferative Assay*

3.6.1. In Vitro Cell Culture

Human breast adenocarcinoma (MCF-7), cervical cancer (HeLa), and liver carcinoma (HepG2) cell lines were purchased from the American Type Culture Collection (ATCC, Alexandria, MN, USA). The tumor cell lines were preserved at the National Cancer Institute, Cairo, Egypt, by serial sub-culturing. The cells were sub-cultured on RPMI 1640 medium supplemented with 1% penicillin/streptomycin and 10% fetal bovine serum [47].

3.6.2. Sulforhodamine B Assay

The antiproliferative activity was determined by the sulforhodamine B (SRB) assay. SRB is a fluorescent aminoanthracene dye with two sulfonic acid groups that bind to the amino groups of intracellular proteins under mildly acidic conditions to provide a sensitive index of cellular protein content. The test was performed as previously described in detail [48,49]. Cells were seeded in 96-well microtiter plates at an initial concentration and left for 24 h to attach to the plates. Then, the three isolated compounds were dissolved in dimethyl sulfoxide (DMSO) and added at different concentrations of 0, 5, 12.5, 25, and 50 µg/mL. After incubation for 48 h, 50 µL of 10% trichloroacetic acid were added and incubated for 60 min at 4 °C to fix the attached cells. The plates were washed and stained with 50 µL of 0.4% SRB dissolved in 1% acetic acid for 30 min at room temperature. The plates were then air-dried, and the dye was solubilized with 100 µL/well of 10 M tris base of pH 10.5 (AppliChem®, Darmstadt, Germany). Optical density (OD) was measured spectrophotometrically at 570 nm using an ELISA microplate reader (Sunrise Tecan reader, Crailsheim, Germany). The experiment was repeated three times and IC_{50} values (concentration that causes 50% decrease in cell viability) were calculated. Doxorubicin served as a positive control at the same concentration range.

4. Conclusions

From the CH_2Cl_2/MeOH extract of the Red Sea grass *T. ciliatum*, a new ergosterol derivative, compound **1**, was isolated. It was identified as 2β,18-dihydroxy-15α-acetoxy-5,6,7,8-tetrahydroergosterol and named thalassosterol. The new compound displayed a significant cytotoxic activity against cervical (HeLa) and breast cancer cell lines (MCF-7), with IC_{50} values of 8.12 and 14.24 µM, respectively. In addition, docking studies with compound **1** explained how it works as an aromatase inhibitor, serving a protocol of cervical and breast cancer treatment. LC-MS based metabolic analysis of *T. ciliatum* furthermore revealed the presence of phenolic compounds, sterols, ceramides, and di- and triglycerides. Therefore, the presented work highlights *T. ciliatum* as an important marine source of bioactive secondary metabolites that should attract further chemical and pharmacological investigation.

Supplementary Materials: The following are available online at http://www.mdpi.com/1660-3397/18/7/354/s1, Figures S1–S2: LC-ESI-HR-MS of the crude extract (positive and negative modes), Figures S3–S10: ESI-HR-MS, 1H NMR, 13C NMR, DEPT-135, HSQC, HMBC, COSY and NOESY of compound 1.

Author Contributions: Conceptualization, U.R.A., S.A.A., R.F.A.A., G.B., H.A.H., A.K.I.; methodology, M.S.G., R.F.A.A., E.S.H.; data curation, R.F.A.A., M.S.G., J.R.F., E.E.E., A.M.A., S.F., A.M.A.E.-k., T.A.-W.; original draft preparation, M.S.G., U.R.A.; writing, review, and editing, all authors. All authors have read and agreed to the published version of the manuscript.

Funding: This publication was funded by the German Research Foundation (DFG) and supported by the Open Access Publication Fund of the University of Wuerzburg.

Acknowledgments: The authors are grateful to Tarek A. Temraz, Marine Science Department, Faculty of Science, Suez Canal University, Egypt, for his taxonomical identification of the Red Sea seagrass *Thalassodendron ciliatum*. S.F. thanks the German Academic Exchange Service (Deutscher Akademischer Austauschdienst, DAAD) for a generous scholarship grant.

Conflicts of Interest: The authors declare no conflict of interest.

References

1. Pergent-Martini, C.; Leoni, V.; Pasqualini, V.; Ardizzone, G.D.; Balestri, E.; Bedini, R.; Belluscio, A.; Belsher, T.; Borg, J.; Boudouresque, C.F.; et al. Descriptors of Posidonia oceanica meadows: Use and application. *Ecol. Indic.* **2005**, *5*, 213–230. [CrossRef]
2. Bandeira, S. Dynamics, biomass and total rhizome length of the seagrass Thalassodendron ciliatum at Inhaca Island, Mozambique. *Plant Ecol.* **1997**, *130*, 133–141. [CrossRef]
3. Short, F.; Carruthers, T.; Dennison, W.; Waycott, M. Global seagrass distribution and diversity: A bioregional model. *J. Exp. Mar. Biol. Ecol.* **2007**, *350*, 3–20. [CrossRef]
4. Kamermans, P.; Hemmingaa, M.A.; Marbàa, N.; Mateoa, M.A.; Mtolerab, M.; Stapel, J. Leaf production, shoot demography, and flowering of Thalassodendron ciliatum along the east African coast. *Aquat. Bot.* **2001**, *70*, 243–258. [CrossRef]
5. Lipkin, Y. Thalassodendron ciliatum in Sinai (northern Red Sea) with special reference to quantitative aspects. *Aquat. Bot.* **1998**, *31*, 125–139. [CrossRef]
6. Hamdy, A.A.; Mettwally, W.S.A.; El Fotouh, M.A.; Rodriguez, B.; El-Dewany, A.I.; El-Toumy, S.A.A.; Hussein, A.A. Bioactive Phenolic Compounds from the Egyptian Red Sea Seagrass Thalassodendron ciliatum. *Z. Naturforsch.* **2012**, *67*, 291–296. [CrossRef] [PubMed]
7. Ibrahim, A.K.; Youssef, A.I.; Arafa, A.S.; Foad, R.; Radwan, M.M.; Ross, S.; Hassanean, H.A.; Ahmed, S.A. Anti-H5N1 virus new diglyceride ester from the Red Sea grass Thallasodendron ciliatum. *Nat. Prod. Res.* **2013**, *27*, 1625–1632. [CrossRef]
8. Ramah, S.; Etwarysing, L.; Auckloo, N.; Gopeechund, A.; Bhagooli, R.; Bahorun, T. Prophylactic antioxidants and phenolics of seagrass and seaweed species: A seasonal variation study in a Southern Indian Ocean Island, Mauritius. *IJMU* **2014**, *9*, 27–37.
9. Mohammed, M.M.D.; Hamdy, A.A.; El-Fiky, N.M.; Mettwally, W.S.A.; El-Beih, A.A.; Kobayashi, N. Anti-influenza A virus activity of a new dihydrochalcone diglycoside isolated from the Egyptian seagrass Thalassodendron ciliatum (Forsk.). *Nat. Prod. Res.* **2014**, *28*, 377–382. [CrossRef]
10. Abdelhameed, R.F.; Ibrahim, A.K.; Yamada, K.; Ahmed, S.A. Cytotoxic and anti-inflammatory compounds from Red Sea grass Thalassodendron ciliatum. *Med. Chem. Res.* **2018**, *27*, 1238–1244. [CrossRef]
11. Goda, M.S.; Eltamany, E.E.; Habib, E.S.; Hassanean, H.A.; Ahmed, A.A.; Abdelhameed, R.F.A.; Ibrahim, A.K. Gas Chromatography-Mass Spectrometry Analysis of Marine Seagrass Thalassodendron ciliatum Collected from Red Sea. *Rec. Pharm. Biomed. Sci.* **2020**, *4*, 1–15. [CrossRef]
12. Sica, D.; Piccialli, V.; Masullo, A. Configuration at C-24 of sterols from the marine phanerogames Posidonia oceanica and Cymodocea nodosa. *Phytochemistry* **1984**, *23*, 2609–2611. [CrossRef]
13. Kontiza, I.; Abatis, D.; Malakate, K.; Vagias, C.; Roussis, V. 3-Keto steroids from the marine organisms Dendrophyllia cornigera and Cymodocea nodosa. *Steroids* **2006**, *71*, 177–181. [CrossRef]
14. Gilian, F.T.; Hogg, R.W.; Drew, E.A. The sterol and fatty acid compositions of seven tropical seagrasses from North Queensland, Australia. *Phytochemistry* **1984**, *23*, 2817–2821. [CrossRef]
15. Lu, W.; Clasquin, M.F.; Melamud, E.; Amador-Noguez, D.; Caudy, A.A.; Rabinowitz, J.D. Metabolomic Analysis via Reversed-Phase Ion-Pairing Liquid Chromatography Coupled to a Stand Alone Orbitrap Mass Spectrometer. *Anal. Chem.* **2010**, *82*, 3212–3221. [CrossRef]
16. Liu, J.; Meng, C.; Liu, S.; Kan, J.; Jin, C. Preparation and characterization of protocatechuic acid grafted chitosan films with antioxidant activity. *Food Hydrocoll.* **2017**, *63*, 457–466. [CrossRef]
17. Cheng, Z.; Kuo, S.; Chan, S.; Ko, F.; Teng, C. Antioxidant properties of butein isolated from Dalbergia odorifera. *Biochim. Biophys. Acta* **1998**, *1392*, 291–299. [CrossRef]
18. Goda, M.S.; Ahmed, S.A.; El Sherif, F.; Hassanean, H.A.; Ibrahim, A.K. Genetically stable plants with boosted flavonoids content after in vitro regeneration of the endangered *Capparis spinosa* L. *Glob. Drugs Ther.* **2017**, *2*, 1–7. [CrossRef]
19. Kirakosyan, A.; Mitchell Seymour, E.; Noon, K.R.; Urcuyo Llanes, D.E.; Kaufman, P.B.; Warber, S.L.; Bolling, S.F. Interactions of antioxidants isolated from tart cherry (Prunus cerasus) fruits. *Food Chem.* **2010**, *122*, 78–83. [CrossRef]
20. Yen, G. Antioxidant activity of anthraquinones and anthrone. *Food Chem.* **2000**, *70*, 437–441. [CrossRef]
21. Wang, Y.; Chan, F.L.; Chen, S.; Leung, L.K. The plant polyphenol butein inhibits testosterone-induced proliferation in breast cancer cells expressing aromatase. *Life Sci.* **2005**, *77*, 39–51. [CrossRef]

22. Asif, M. Health effects of omega- 3,6,9 fatty acids: Perilla frutescens is a good example of plant oils. *Orient. Pharm. Exp. Med.* **2011**, *11*, 51–59. [CrossRef] [PubMed]

23. Kilic, T.; Carikci, S.; Topcu, G.; .Aslan, I.; Goren, A.C. Diterpenoids from Sideritis condensata. Evaluation of chemotaxonomy of Sideritis species and insecticidal activity. *Chem. Nat. Compd.* **2009**, *45*, 918–920. [CrossRef]

24. Smyrniotopoulos, V.; Vagias, C.; Bruyère, C.; Lamoral-Theys, D.; Kiss, R.; Roussis, V. Structure and in vitro antitumor activity evaluation of brominated diterpenes from the red alga Sphaerococcus coronopifolius. *Bioorg. Med. Chem.* **2010**, *18*, 1321–1330. [CrossRef] [PubMed]

25. Chen, S.; Yong, T.; Zhang, Y.; Su, J.; Jiao, C.; Xie, Y. Anti-tumor and anti-angiogenic ergosterols from Ganoderma lucidum. *Front. Chem.* **2017**, *5*, 85. [CrossRef]

26. Osman, M.; Sazalli, N.N.H.; Willium, D.; Golam, F.; Nezhadahmadi, A. Potentiality of Roselle and On-ion (Allium cepa) peel as Raw Materials for Producing Protocatechuic Acid in Tropical Malaysia: A Comparative Study. *INDJST* **2014**, *7*, 1847–1851.

27. Sebolai, O.M.; Kock, J.L.F.; Pohl, C.H.; Botes, P.J.; Nigam, S. Report on the discovery of a novel 3-hydroxy oxylipin cascade in the yeast Saccharomycopsis synnaedendra. *Prostaglandins Other Lipid Mediat.* **2004**, *74*, 139–146. [CrossRef]

28. Mikhailova, M.V.; Bemis, D.L.; Wise, M.L.; Gerwick, W.H.; Norris, J.N.; Jacobs, R.S. Structure and biosynthesis of novel conjugated polyene fatty acids from the marine green alga Anadyomene stellate. *Lipids* **1995**, *30*, 583–589. [CrossRef]

29. Zhang, Y.; Seeram, N.P.; Lee, R.; Feng, L.; Heber, D. The Isolation and Identification of Isolation and Identification of Strawberry Phenolics with Antioxidant and Human Cancer Cell Antiproliferative Properties. *J. Agric. Food Chem.* **2008**, *56*, 670–675. [CrossRef]

30. Wang, S.; Li, X.M.; Teuscher, F.; Li, D.L.; Diesel, A.; Ebel, R.; Proksch, P.; Wang, B.G. Chaetopyranin, a benzaldehyde derivative, and other related metabolites from Chaetomium globosum, an endophytic fungus derived from the marine red alga Polysiphonia urceolata. *J. Nat. Prod.* **2006**, *69*, 1622–1625. [CrossRef]

31. Paranjape, S.R.; Chiang, Y.M.; Sanchez, J.F.; Entwistle, R.; Wang, C.C.; Oakley, B.R.; Gamblin, T.C. Inhibition of Tau aggregation by three Aspergillus nidulans secondary metabolites: 2,ω-dihydroxyemodin, asperthecin, and asperbenzaldehyde. *Planta Med.* **2014**, *80*, 77–85. [CrossRef] [PubMed]

32. Cao, X.; Wei, Y.; Ito, Y. Preparative Isolation of Isorhamnetin from Stigma maydis using High Speed Countercurrent Chromatography. *J. Liq. Chromatogr. Relat. Technol.* **2008**, *32*, 273–280. [CrossRef] [PubMed]

33. Smyrniotopoulos, V.; Vagias, C.; Roussis, V. Sphaeroane and Neodolabellane Diterpenes from the Red Alga Sphaerococcus coronopifolius. *Mar. Drugs* **2009**, *7*, 184–195. [CrossRef] [PubMed]

34. Nieto, I.; Avila, I. Determination of fatty acids and triterpenoid compounds from the fruiting body of Suillus luteus. *Rev. Colomb. Quim.* **2008**, *37*, 297–304.

35. Zhang, W.; Che, C. Isomalabaricane-Type Nortriterpenoids and Other Constituents of the Marine Sponge Geodia japonica. *J. Nat. Prod.* **2001**, *64*, 1489–1492. [CrossRef]

36. Liu, Y. *Dietary Chinese Herbs*, 1st ed.; Springer: Vienna, Austria, 2015; pp. 789–796. [CrossRef]

37. Goad, L.J.; Akihisa, T. *Analysis of Sterols*, 1st ed.; Blackie Academic and Professional: London, UK, 1997; pp. 1–437. [CrossRef]

38. Ványolòs, A.; Béni, Z.; Dékány, M.; Simon, A.; Báthori, M. Novel Ecdysteroids from Serratula wolffii. *Sci. World J.* **2012**, 1–5. [CrossRef]

39. Shady, N.H.; Abdelmohsen, U.R.; Ahmed, S.; Fouad, M.; Kamel, M.S. Phytochemical and biological investigation of the red sea marine sponge Hyrtios sp. *J. Pharmacogn. Phytochem.* **2017**, *6*, 241–246.

40. Ghosh, T.; Maity, T.K.; Singh, J. Evaluation of antitumor activity of stigmasterol, a constituent isolated from Bacopa monnieri Linn. aerial parts against Ehrlich Ascites Carcinoma in mice. *Orient. Pharm. Exp. Med.* **2011**, *11*, 41–49. [CrossRef]

41. Roleira, F.M.F.; Varela, C.; Amaral, C.; Costa, S.C.; Correia-da-Silva, G.; Moraca, F.; Costa, G.; Alcaro, S.; Teixeira, N.A.A.; Tavares da Silva, E.J. C-6α- vs C-7α-Substituted Steroidal Aromatase Inhibitors: Which Is Better? Synthesis, Biochemical Evaluation, Docking Studies, and Structure–Activity Relationships. *J. Med. Chem.* **2019**, *62*, 3636–3657. [CrossRef]

42. Rampogu, S.; Ravinder, D.; Pawar, S.C.; Lee, K.W. Natural Compound Modulates the Cervical Cancer Microenvironment—A Pharmacophore Guided Molecular Modelling Approaches. *J. Clin. Med.* **2018**, *7*, 551. [CrossRef] [PubMed]

43. Kumar, S.; Bajaj, S.; Bodla, R.B. Preclinical screening methods in cancer. *Indian J. Pharmacol.* **2016**, *48*, 481–486. [CrossRef] [PubMed]

44. Ibrahim, R.S.; Seif El-Din, A.A.; Abu-Serie, M.; Abd El Rahman, N.M.; El-Demellawy, M.; Metwally, A.M. Investigation of In-Vitro Cytotoxic and Potential Anticancer Activities of Flavonoidal Aglycones from Egyptian Propolis. *Rec. Pharm. Biomed. Sci.* **2017**, *2*, 13–20. [CrossRef]

45. Abdelhafez, O.H.; Othman, E.M.; Fahim, J.R.; Desouky, S.Y.; Pimentel-Elardo, S.M.; Nodwell, J.R.; Schirmeister, T.; Tawfike, A.; Abdelmohsen, U.R. Metabolomics analysis and biological investigation of three Malvaceae plants. *Phytochem. Anal.* **2020**, *31*, 204–214. [CrossRef]

46. Gasteiger, J.; Rudolph, C.; Sadowski, J. Automatic generation of 3D-atomic coordinates for organic molecules. *Tetrahedron Comput. Methodol.* **1990**, *3*, 537–547. [CrossRef]

47. Skehan, P.; Strong, R.; Scudiero, D.; Monks, A.; McMahon, J.; Vistica, D.; Warren, J.T.; Bokesch, H.; Kenney, S.; Boyd, M.R. New colorimetric cytotoxicity assay for anticancer drug screening. *J. Natl. Cancer Inst.* **1990**, *82*, 1107–1112. [CrossRef] [PubMed]

48. Vichai, V.; Kirtikara, K. Sulforhodamine B colorimetric assay for cytotoxicity screening. *Nat. Protoc.* **2006**, *1*, 1112–1116. [CrossRef]

49. Abu-Dahab, R.; Afifi, F. Antiproliferative activity of selected medicinal plants of Jordan against a breast adenocarcinoma cell line (MCF7). *Sci. Pharm.* **2007**, *75*, 121–136. [CrossRef]

 *marine drugs*

Article

Induction of Antibacterial Metabolites by Co-Cultivation of Two Red-Sea-Sponge-Associated Actinomycetes *Micromonospora* sp. UR56 and *Actinokinespora* sp. EG49

Mohamed S. Hifnawy [1,†], Hossam M. Hassan [2,†], Rabab Mohammed [2], Mohamed M. Fouda [3], Ahmed M. Sayed [3], Ahmed A. Hamed [4], Sameh F. AbouZid [2], Mostafa E. Rateb [5], Hani A. Alhadrami [6,7,*] and Usama Ramadan Abdelmohsen [8,9,*]

[1] Department of Pharmacognosy, Faculty of Pharmacy, Cairo University, 11787 Cairo, Egypt; mhifnawy@hotmail.com
[2] Department of Pharmacognosy, Faculty of Pharmacy, Beni-Suef University, 62514 Beni-Suef, Egypt; abuh20050@yahoo.com (H.M.H.); rmwork06@yahoo.com (R.M.); sameh.zaid@pharm.bsu.edu.eg (S.F.A.)
[3] Department of Pharmacognosy, Faculty of Pharmacy, Nahda University, 62513 Beni-Suef, Egypt; foudapharma@gmail.com (M.M.F.); ahmed.mohamed.sayed@nub.edu.eg (A.M.S.)
[4] Microbial Chemistry Department, National Research Center, 33 El-Buhouth Street, 12622 Giza, Egypt; ahmedshalbio@gmail.com
[5] School of Computing, Engineering & Physical Sciences, University of the West of Scotland, Paisley PA1 2BE, UK; mostafa.rateb@uws.ac.uk
[6] Department of Medical Laboratory Technology, Faculty of Applied Medical Sciences, King Abdulaziz University, Jeddah 21589, Saudi Arabia
[7] Special Infectious Agent Unit, King Fahd Medical Research Centre, King Abdulaziz University, Jeddah 21589, Saudi Arabia
[8] Department of Pharmacognosy, Faculty of Pharmacy, Minia University, 61519 Minia, Egypt
[9] Department of Pharmacognosy, Faculty of Pharmacy, Deraya University, Universities Zone, P.O. Box 61111 New Minia City, Egypt
* Correspondence: hanialhadrami@kau.edu.sa (H.A.A.); usama.ramadan@mu.edu.eg (U.R.A.)
† These authors equally contributed to this work.

Received: 27 March 2020; Accepted: 30 April 2020; Published: 5 May 2020

Abstract: Liquid chromatography coupled with high resolution mass spectrometry (LC-HRESMS)-assisted metabolomic profiling of two sponge-associated actinomycetes, *Micromonospora* sp. UR56 and *Actinokineospora* sp. EG49, revealed that the co-culture of these two actinomycetes induced the accumulation of metabolites that were not traced in their axenic cultures. Dereplication suggested that phenazine-derived compounds were the main induced metabolites. Hence, following large-scale co-fermentation, the major induced metabolites were isolated and structurally characterized as the already known dimethyl phenazine-1,6-dicarboxylate (1), phenazine-1,6-dicarboxylic acid mono methyl ester (phencomycin; 2), phenazine-1-carboxylic acid (tubermycin; 3), N-(2-hydroxyphenyl)-acetamide (9), and *p*-anisamide (10). Subsequently, the antibacterial, antibiofilm, and cytotoxic properties of these metabolites (1–3, 9, and 10) were determined in vitro. All the tested compounds except 9 showed high to moderate antibacterial and antibiofilm activities, whereas their cytotoxic effects were modest. Testing against *Staphylococcus* DNA gyrase-B and pyruvate kinase as possible molecular targets together with binding mode studies showed that compounds 1–3 could exert their bacterial inhibitory activities through the inhibition of both enzymes. Moreover, their structural differences, particularly the substitution at C-1 and C-6, played a crucial role in the determination of their inhibitory spectra and potency. In conclusion, the present study highlighted that microbial co-cultivation is an efficient tool for the discovery of new antimicrobial candidates and indicated phenazines as potential lead compounds for further development as antibiotic scaffold.

Keywords: co-cultivation; phenazine; sponge-associated actinomycetes; antibacterial; antibiofilm; DNA gyrase; pyruvate kinase

---

1. Introduction

Natural products derived from marine microbes play a pivotal role in drug discovery and development due to their diverse molecular and chemical scaffolds, which cannot be matched by any synthetic or combinatorial libraries [1]. Marine-sponge-associated actinomycetes have recently been used to produce a multitude of new bioactive compounds with novel molecular scaffolds and potent pharmacological activities such as antibacterial, antifungal, antiparasitic, immunomodulatory, anti-inflammatory, antioxidant, and anticancer activities [2–5]. However, searching for new and promising bioactive secondary metabolites from marine microbes is becoming a serious challenge due to the increasing rate of rediscovery of known secondary metabolites [6,7]. On the other hand, genomic work has revealed that definite groups of bacteria and fungi have huge numbers of silent biosynthetic gene clusters (BGCs) that encode for secondary metabolites that are not traced under normal laboratory conditions [8,9]. Microbial competition for space and nutrition are considered eventual routes for the induction of bioactive metabolites, mainly antimicrobial agents [10]. Several protocols have been used to trigger such cryptic biosynthetic pathways. One of these protocols involves co-cultivation or the so-called mixed fermentation of two or more microorganisms in a single culture flask, which results in the production of antimicrobial secondary metabolites [11–13]. Recently, we reported the production of new antitumor agents, saccharomonosporine A and convolutamydine F, by the co-fermentation of two marine-sponge-associated actinomycetes, *Dietzia* sp. UR66 and *Saccharomonospora* sp. UR22, obtained from *Callyspongia siphonella* [14]. A chlorinated benzophenone pestalone that showed potent antibiotic activity was sourced from the co-cultivation of two marine-associated fungi, α-proteobacterium CNJ-328 and *Pestalotia* sp. CNL-365 [15]. The induction of cryptic pulicatin derivatives that exhibit potent antifungal effects through the microbial co-culture of *Pantoea agglomerans* with *Penicillium citrinum* was recently reported [16]. Finally, the induced production of emericellamides A and B obtained from a co-fermentation of the marine-associated fungus *Emericella* sp. CNL-878 and the marine derived bacterium *Salinispora arenicola* was reported [17].

Phenazine compounds are heterocyclic nitrogenous compounds that consist of two benzene rings attached through two nitrogen atoms and substituted at different sites of the core ring system. Phenazine derivatives have been found to show a wide range of biological activities, including antibacterial, antiviral, antitumor, antimalarial, and antiparasitic activities [18,19]. They have been isolated in large amounts from terrestrial bacteria such as *Pseudomonas*, *Streptomyces*, and other genera from marine habitats [20]. The iminophenazine clofazimine is an example of a successful phenazine derivative, having been approved by the FDA for the treatment of leprosy and drug-resistant *Mycobacterium tuberculosis* strains [21,22]. Another example of a phenazine is bis-(phenazine-1-carboxamide), which acts as a strong cytotoxin and represents an attractive class of anticancer drugs [23]. In an earlier work, we found that *Actinokineospora* sp. EG49 was able to induce *Nocardiopsis* sp. RV163 to produce 1,6-dihydroxyphenazine upon co-cultivation [24]. On the other hand, *Micromonospora* sp. are widespread actinomycetes and prolific producers of diverse antibiotics [25,26]. Consequently, we decided to extend our co-cultivation trials on both marine-derived *Actinokineospora* sp. EG49 and *Micromonospora* sp. UR56 in order to induce the production of further antibacterial metabolites, which were also found to be of the phenazine class. Based on earlier reports on the biological activities of this class of compounds, we suggested both DNA gyrase B (Gyr-B) and pyruvate kinase (PK) to be the possible molecular targets of their antibacterial activity. The working outline of the present study is illustrated in Figure 1.

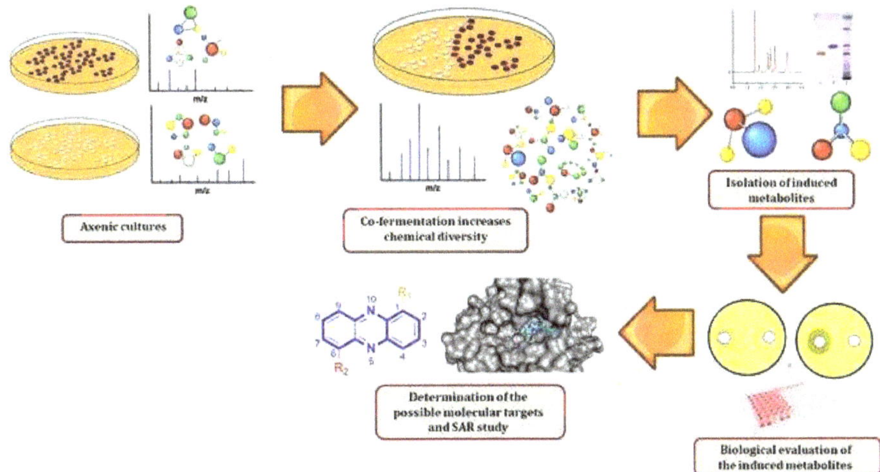

Figure 1. Outline of the procedure used in the present study.

2. Results and Discussion

2.1. Metabolomic Profiles of the Axenic and Co-Culture Extracts

The chemical profiles of the actinomycetes *Micromonospora* sp. UR56 and *Actinokineospora* sp. EG49 were investigated via liquid chromatography coupled with mass spectrometry (LC-HRMS) analysis after their fermentations (axenic and co-fermentation). The metabolomic profile of the co-culture extract displayed the induction of diverse metabolites from different chemical classes compared to those of the two axenic cultures (Figure 2, Supplementary Figure S32, and Supplementary Table S3). Twelve metabolites were putatively identified in the *Micromonospora* sp. UR56-derived extract, where phenazine derivatives were found to prevail (Figure 2; Figure 3, Supplementary Figure S30). Most of these dereplicated phenazines e.g., phenazine-1-carboxylic acid (**3**), aestivophoenin c (**8**), and methyl saphenate (**4**) have been reported to possess antimicrobial and cytotoxic properties [27]. The remaining identified compounds were found to belong to the N-containing and polyketide classes. Within the axenic *Actinokineospora* sp. EG49 culture, no phenazine derivatives were traced in the LC-HRMS analysis of the extract. Additionally, its chemical profile revealed poor diversity, with a few identified N-containing and polyketide metabolites (Supplementary Figure S31 and Supplementary Table S2). On the other hand, the mixed fermentation of both actinomycetes induced *Micromonospora* sp. UR56 to accumulate diverse phenazine derivatives (**1–8**) (Figure 2). Such induction could be due to environmental competition or chemical defense mechanisms [10]. Based on the metabolomic profiling of the co-culture, the major induced metabolites (**1–3**, **9**, and **10**) were targeted and isolated using Sephadex LH20 followed by silica gel column chromatography, and identified using different spectroscopic approaches. Subsequently, they were subjected to antibacterial, antibiofilm, and cytotoxicity testing.

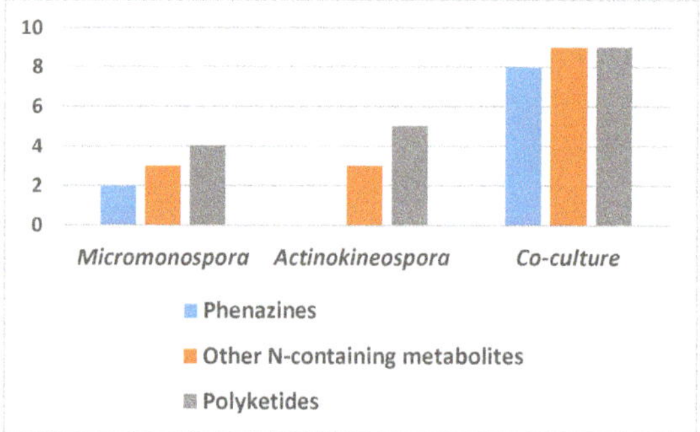

Figure 2. Classes of metabolites produced from *Micromonospora* sp. UR56 and *Actinokineospora* sp. EG49 axenic and co-cultures.

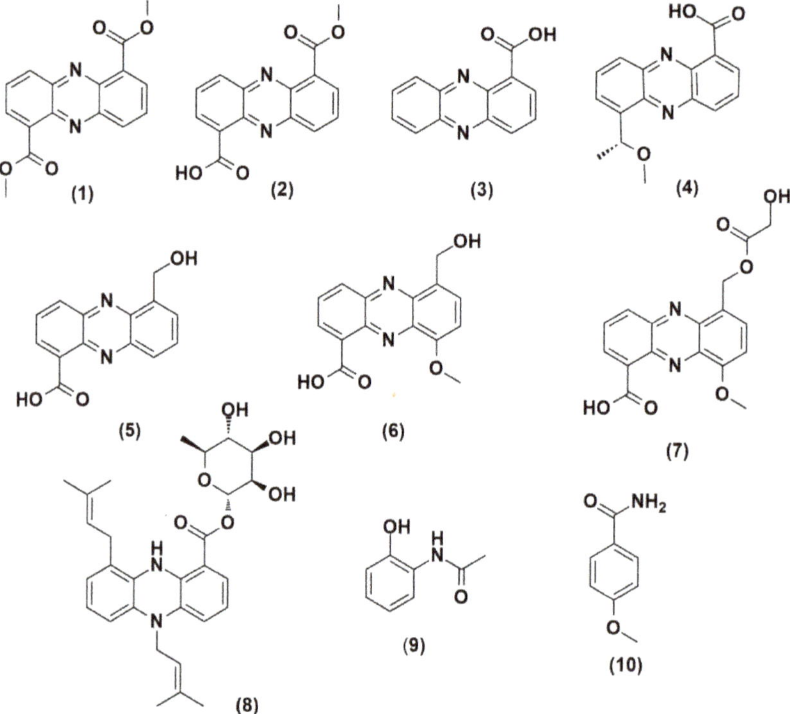

Figure 3. Identified phenazine derivatives in the axenic *Micromonospora* sp. UR56 culture, and after its co-culture with *Actinokineospora* sp. EG49. **1**: dimethyl phenazine-1,6-dicarboxylate, **2**: phencomycin, **3**: phenazine-1-carboxylic acid, **4**: methyl saphenate, **5**: 1-hydroxy methyl-6-carboxy phenazine, **6**: griseolutic acid, **7**: griseolutin A, **8**: aestivophoenin C, **9**: N-(2-hydroxyphenyl)-acetamide, and **10**: *p*-anisamide.

2.2. Bioactivity Testing

2.2.1. Antibacterial Activity

Based on earlier reports on their antibacterial potential, phenazine-derived compounds have proven to be an interesting chemical scaffold for the development of new antibacterial agents [27]. Therefore, all isolated compounds were evaluated against *Staphylococcus aureus* ATCC9144, *Bacillus subtilis* ATCC29212, *Pseudomonas aeruginosa* ATCC27853, and *Escherichia coli* ATCC25922 (Table 1). Compounds **3** and **10** displayed potent antibacterial activity against *P. aeruginosa* with growth inhibition of 94% and 70%, respectively, while compounds **1**, **2**, and **9** showed considerable antibacterial activity against *S. aureus* with growth inhibition of 47%, 69%, and 53%, respectively (Table 1). Based on these results together those previously reported [27], we concluded that the phenazine-1-carboxylic acid scaffold is essential for antibacterial activity against Gram-negative bacteria and produces no observable activity towards Gram-positive ones. Although the addition of another carboxylic acid or carboxyl ester at C-6 significantly decreased the inhibitory activity against Gram-negative bacteria, it converted these phenazine derivatives to be active against Gram-positive strains (Figure 4).

Table 1. Growth inhibition % of **1–3**, **9**, and **10** at 15 μM against different bacterial strains.

Tested Compounds	*Staphylococcus aureus*	*Bacillus subtilis*	*Escherichia coli*	*Pseudomonas aeruginosa*
1	57.19 ± 1.2	19.61 ± 1.5	9.93 ± 1.3	23.41 ± 1.7
2	68.87 ± 0.8	32.13 ± 1.3	11.15 ± 1.4	23.25 ± 2.2
3	NA	1.69 ± 1.2	24.34 ± 1.8	94.17 ± 2.8
9	1.1 ± 0.9	NA	NA	4.74 ± 1.6
10	53.17 ± 1.2	42.16 ± 1.9	19.27 ± 2.5	70.23 ± 1.1
Gentamicin	99.1 ± 0.7	97 ± 1.6	99 ± 0.6	99.7 ± 0.2

NA: no activity.

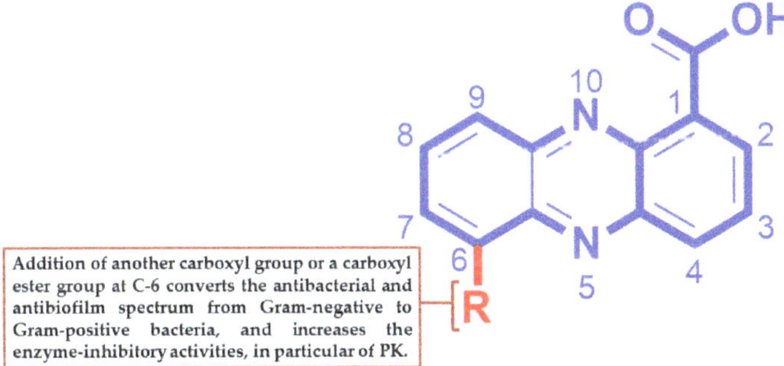

Figure 4. Structure–activity relationship of the induced phenazine derivatives.

2.2.2. Antibiofilm Activity

To further evaluate the inhibitory activities of the isolated metabolites (**1–3**, **9**, and **10**), their antibiofilm potentials against *S. aureus*, *B. subtilis*, *E. coli*, and *P. aeruginosa* were determined. Compounds **3** and **10** displayed potent antibiofilm activity against *P. aeruginosa* with % inhibition rates of 94% and 73%, respectively, while compounds **1**, **2**, and **9** showed mild to moderate inhibitory activity against *E. coli* with % inhibition ranges of 34–54%. Compounds **1** and **2** both showed potent to moderate inhibitory activity against *S. aureus* with % inhibition rates of 50% and 75%, respectively (Table 2). Similar to the antibacterial results, the presence of carboxylic acids on both C-1 and C-6 of the phenazine ring system decreased the antibiofilm effect towards Gram-negative strains, but made these derivatives active against Gram-positive ones, particularly, *S. aureus* (Figure 4).

2.2.3. Cytotoxic Activity

The five induced compounds (**1–3**, **9**, and **10**) were additionally tested for their cytotoxic activity against four human cancer cell lines (Table 3). Only compound **9** was able to induce moderate cytotoxicity towards the tested cell lines, with IC_{50} values ranging from 10 to 36 μM. Regarding the activity of phenazine derivatives (**1–3**), compound **2** was found to be the most active against all tested cell lines. Previous studies have suggested that phenazine-related compounds can exert significant cytotoxic activities through the inhibition of topoisomerase enzymes [28–30].

Table 2. Antibiofilm activity of **1–3**, **9**, and **10** at 2 μM against different bacterial strains.

Tested Compounds	*S. aureus*	*B. subtilis*	*E. coli*	*P. aeruginosa*
1	50.26 ± 0.4	12.10 ± 3.2	36.66 ± 2.9	18.36 ± 0.9
2	75.10 ± 2.4	18.65 ± 1.6	54.07 ± 2.5	22.28 ± 1.5
3	NA	NA	54.67 ± 1.4	93.98 ± 2.2
9	11.47 ± 2.9	4.91 ± 1.8	34.55 ± 2.6	7.39 ± 1.9
10	61.20 ± 3.7	20.29 ± 1.1	57.47 ± 3.1	73.52 ± 1.3

NA: no activity.

Table 3. Cytotoxic activity of isolated compounds **1–3**, **9**, and **10** at 2 μM.

Tested Compounds	WI38	HCT116	HePG-2	MCF7
1	63.18 ± 3.6	85.04 ± 3.9	92.06 ± 4.7	>100
2	76.30 ± 3.9	60.81 ± 3.5	76.11 ± 3.9	82.24 ± 4.4
3	51.22 ± 3.2	91.27 ± 4.6	>100	>100
9	36.47 ± 2.3	14.56 ± 1.2	10.16 ± 0.9	12.65 ± 1.1
10	>100	>100	>100	>100
Doxorubicin	6.72 ± 0.5	5.23 ± 0.3	4.50 ± 0.2	4.17±0.2

2.2.4. In Vitro Enzyme Assay

DNA gyrase is a topoisomerase-type enzyme present exclusively in bacterial cells. Inhibition of its subunit A (Gyr-A) leads to cell death by trapping the gyrase–DNA complex and preventing DNA replication. On the other hand, inhibition of Gyr-B blocks ATPase activity and thus deprives the cell of the energy source needed for DNA replication [31]. Gyr-B as a molecular target offers an opportunity to avoid cross-resistance to the well-known Gyr-A inhibitors, quinolones. Several phenazine and acridine derivatives have been reported to be topoisomerase I and II inhibitors in human cancer cells [28–30]. The characteristic planar structure of this class of compounds directs them to interact with the ATP-binding domains of the enzymes (ATPase pocket) [28]. On the other hand, PK is known to be a critical enzyme in catalyzing the final step of glycolysis, which involves the transfer of a phosphoryl group from phosphoenolpyruvate to ADP, producing pyruvate and ATP [32]. Recently, it was identified as a "superhub", listed among the top 1% of all interacting proteins in *S. aureus* and found to be a key regulator of the quorum-sensing system and the biofilm formation process in staphylococci [33–35]. Considering these points together with the structural similarity of our compounds (**1–3**) to previously well-known topoisomerase inhibitors, we selected *Staphylococcus* DNA gyrase-B and PK as possible molecular targets to mediate the observed antibacterial and antibiofilm activities of the induced phenazine derivatives toward *S. aureus*. In vitro studies (Table 4) showed **1** to be the most active compound against both Gyr-B and PK, followed by **2** and **3**. These finding indicate a direct link between the substitutions on C-1 and C-6 and the Gyr-B- and PK-inhibitory activities of this class of metabolites (Figure 3). Although the enzyme-inhibitory activity of compounds **1–3** was convergent, and compound **1** was slightly more active than compound **2** in the enzyme assay studies, compound **3** was inactive, and **1** was less active than **2** in both inhibitory assays against *S. aureus*. These observations could be attributed to the bacterial cell wall permeability, where *S. aureus* may permit the diffusion of **2** more easily than **1** and prevent the crossing of compound **1**. In the same manner, *P. aeruginosa*'s outer membrane may be selective only for compound **3**, and hence make both other phenazines (**1** and **2**) inactive. According to a previous report [36], small, moderately lipophilic compounds can cross the Gram-positive bacterial membrane more easily. In contrast, Gram-negative bacteria, which have porins (hydrophilic channels) in their outer membranes, are more selective for hydrophilic compounds. The calculated LogP (cLogP) of each compound revealed that compound **2** had the optimum lipophilicity (cLogP = 1.9), whereas compound **1** was less lipophilic (cLogP = 1.12) and compound **3** was more lipophilic (cLogP = 2.97) Such differences in lipophilicity between compounds **1** to **3** could explain the opposite observations between enzyme inhibition and antibacterial and antibiofilm activities of compounds **1–3**.

Table 4. Calculated interactions and affinities of compounds **1–3** with active-site residues of the *Staphylococcus* Gyr-B and PK.

Protein Target	Ligand	IC$_{50}$ (µM) *	Binding Energy (kcal/mol)	Hydrogen Bonding Interactions	Hydrophobic Interactions
Gyr-B	1	19.18 ± 1.69	−7.6	ASN-54, GLU-58	ILE-51, VAL-79 PRO-87, ILE-86 ILE-102, ILE-103 ILE-175
	2	21.28 ± 2.36	−7.5	SER-55, ILE-51	ILE-51, VAL-79 PRO-87, ILE-86 ILE-102, ILE-103 ILE-175
	3	27.69 ± 1.08	−7.2	SER-55, ILE-51	ILE-51, VAL-79 PRO-87, ILE-86 ILE-102, ILE-103 ILE-175
	Co-crystallized ligand	0.091	-	ASP-81, ASN-54	ILE-51, VAL-79 PRO-87, ILE-86 ILE-102, ILE-103 ILE-175
PK	1	7.2 ± 0.07	−7.7	SER-362A, SER-362B THR-366A, THR-353B, ASN-369A	ALA-358B, ILE-361B, LEU-370A
	2	9.3 ± 0.03	−7.5	SER-362A, THR-366A, THR-353B, ASN-369A	ALA-358B, ILE-361B, LEU-370A
	3	22.5 ± 0.04	−6.5	ASN-369B, THR-353A	ALA-358A, ILE-361A
	Co-crystallized ligand	0.24	-	SER-362A, SER-362B, ASN-369B, HIS-365A	ALA-358A, ALA-358B ILE-361B, LEU-370A

* IC$_{50}$ ± SD (µM).

2.3. Docking Study

The potential binding modes of **1–3** with DNA Gyr-B and pyruvate kinase were investigated by docking at their active binding sites. *Staphylococcus* Gyr-B and PK of PDB code (3g7b and 3T0T) were selected for docking tests since they had optimum resolutions (2.3 Å and 3.1 Å) and are co-crystallized with their inhibitors. The spheres surrounding the co-crystallized inhibitors were selected as active sites for docking. All isolated compounds exhibited convergent docking poses (Figure 5) and were comparable to the co-crystallized Gyr-B inhibitor [37]. The oxygens of both ester groups of **1** on C1 and C6 were hydrogen-bonded to ASN-54, similarly to the co-crystallized Gyr-B ligand, and GLU-58. These interactions meant that the whole molecule was well-embedded inside the active site pocket, where it was further stabilized by the hydrophobic interactions between the molecule's aromatic rings and ILE-51, VAL-79, PRO-87, ILE-86, ILE-102, ILE-103, and ILE-175. The absence of an ester group in **2** and **3** led them to take a different and less stable orientation inside the binding pocket and hence, their carboxylic acid moieties interacted from only one side with different amino acid residues via hydrogen bonds (SER-55 and ILE-51); however, they showed similar hydrophobic interactions to **1**. Unlike the co-crystallized Gyr-B inhibitor, **1–3** did not show any interactions with ASP-81. On the other hand, **1** and **2** showed similar binding modes in the PK active site, although different from that of **3**. The co-crystallized PK inhibitor [38] fitted perfectly inside the pocket formed between the enzyme's A and B subunits (Figure 6). This binding mode was stabilized by six main hydrogen bonds with SER-362A, SER-362B, ASN-369B, and HIS-365A (A and B notations correspond to PK subunits). These interacting residues are of particular interest as they are not conserved in human PK, and thus are likely to further influence the selectivity of the bacterial PK inhibitors. Herein, **1** showed a convergent binding mode to the co-crystallized PK inhibitor, where it anchored on one side of the PK active site (Figure 6) through three hydrogen bonds with SER-362A and SER-362B, similarly to the co-crystallized ligand, and through an additional hydrogen bond with THR-366A. All these interactions were from the ester moiety of one side of the molecule; the second ester group on the other side increased the molecular stability inside the binding site through hydrogen bonding to THR-353B and ASN-369A. Additionally, the aromatic planar structure of **1** was sandwiched between two hydrophobic surfaces inside the binding pocket, where it interacted via hydrophobic interactions with ALA-358B and ILE-361B from one side, and with LEU-370A from the other side. Regarding compound **2**, it revealed a binding mode and interactions quite similar to that of **1**. Such strong interactions with the PK binding side explain the good in vitro inhibitory activity of both **1** and **2** (Table 4) in comparison to **3**, which showed different binding modes inside the enzyme's binding cavity. Unlike **1** and **2**, and due to its single carboxylic acid moiety, **3** fitted inside the other corner of the active pocket through only three hydrogen bonds with ASN-369B and THR-353A. Regarding hydrophobic interactions, it interacted with only two residues, ALA-358A, and ILE-361A (Figure 6). These weaker interactions explain the inferior in vitro inhibitory activity of **3** toward PK (Table 4). In addition, it indicates the importance of the 1,6-dicarboxylic acid moieties in the future development of *Staphylococcus* PK inhibitors.

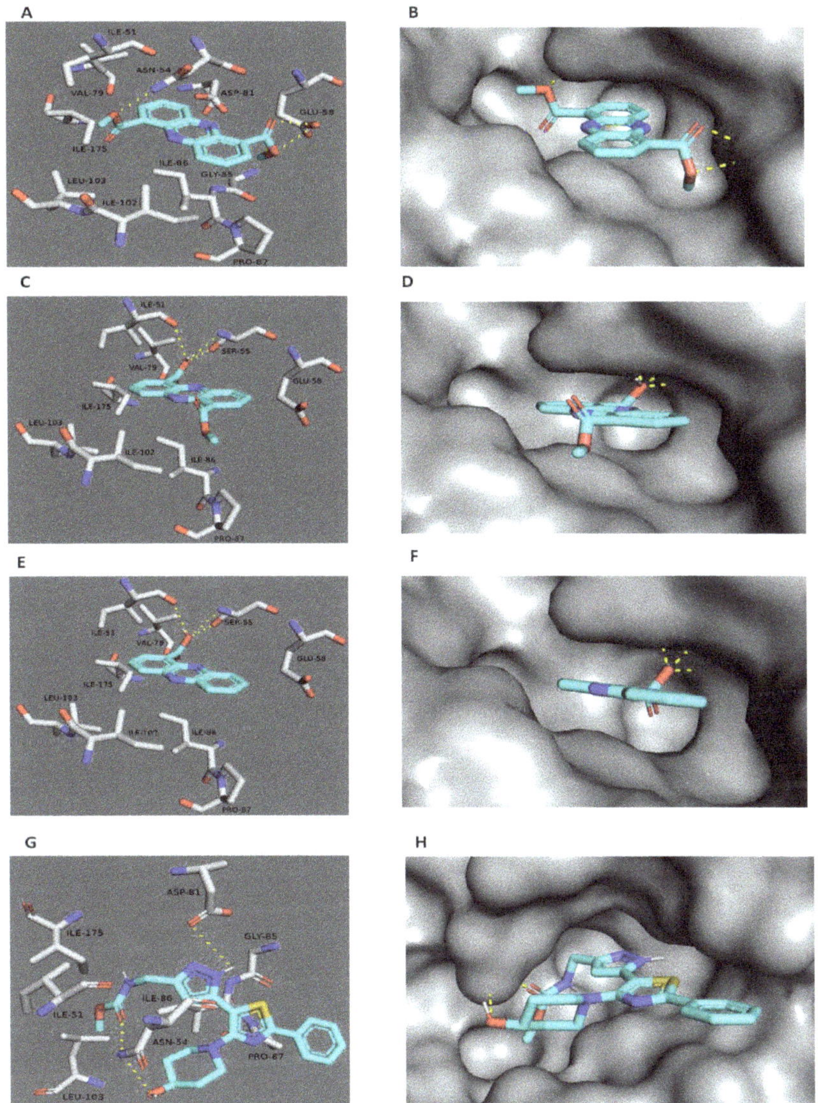

Figure 5. Docking of **1** (**A,B**), **2** (**C,D**) and **3** (**E,F**) within the active site of Staphylococcal Gyr-B. (**G,H**) The key binding interactions of Gyr-B co-crystallized ligand. The amino acid side chains were depicted in (**A,C,E,G**) for clarification.

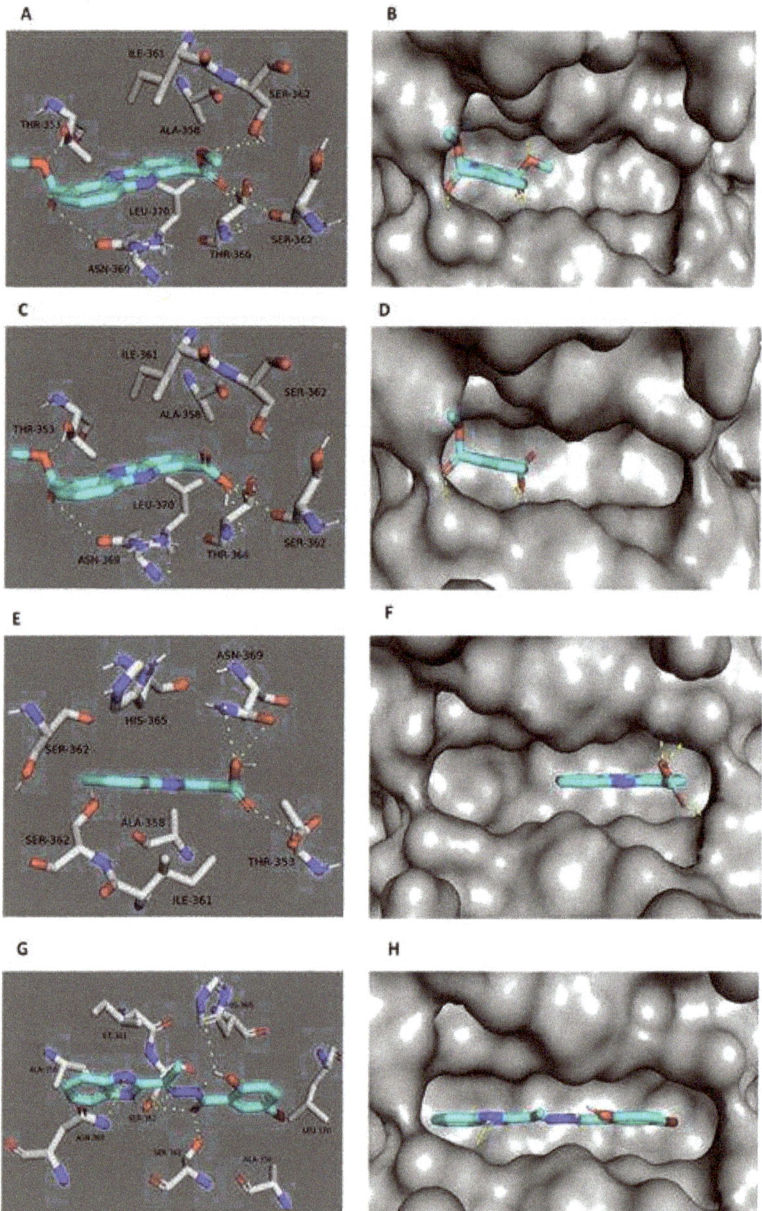

Figure 6. Docking of **1** (**A,B**), **2** (**C,D**), and **3** (**E,F**) within the active site of *Staphylococcus* PK. (**G,H**) The key binding interactions of the Gyr-B co-crystallized ligand. The amino acid side chains are depicted in (**A,C,E,G**) for clarification.

3. Material and Methods

3.1. General Experimental Procedures

The chemical solvents used in this study, such as n-hexane, dichloromethane, ethyl acetate, and methanol, were obtained from Sigma-Aldrich, Saint Louis, Missouri, USA. Silica gel 60 (63–200 μM,

E. Merck, Sigma-Aldrich) and Sephadex LH20 (0.25–0.1 mm, GE Healthcare, Sigma-Aldrich) were used for chromatographic isolation and purification. Thin-layer chromatography was performed using pre-coated silica gel aluminum plates (E. Merck, Darmstadt, Germany, Kieselgel 60 F254, 20 × 20 cm, 0.25 mm). *p*-anisaldehyde (0.5 : 85 : 10 : 5 *p-anisaldehyde* : methanol : glacial acetic acid : sulfuric acid) was used as visualizing spray reagent for different spots accompanied by heating at 110 °C. 1D, 2D, and ^{13}C NMR spectra were recorded on a JEOL ECA-600 spectrometer (600 MHz for ^{1}H and 150 MHz for ^{13}C, respectively). Each sample was dissolved in prober deuterated solvent such as CDCl$_3$ and CD$_3$OD. All of the chemical shifts were recorded and expressed in ppm units related to the TMS signal as an internal standard, and coupling constants (J) were recorded in Hz.

3.2. Sponge Collection

Callyspongia sp. and *Spheciosponge vagabunda* were collected from the Red Sea (Ras Mohamed, Sinai; (GPS coordinates 27°47.655 N; 34°12.904 W) at a depth of 10 m in August 2006. The collected sponges, identified by R.W.M. van Soest (University of Amsterdam, Netherlands), were transferred to plastic bags containing sterile seawater and transported to the laboratory. Sponge biomass was cleaned with sterile seawater, cut into fragments of ca. 1 cm^3, and then carefully homogenized with 10 volumes of seawater. The freshly prepared supernatant was diluted in 10-fold series (10^{-1}, 10^{-2}, 10^{-3}) and subsequently plated out onto agar plates.

3.3. Actinomycetes Isolation

Different media were used for isolation of actinomycetes M1 [39]—ISP2 medium [40], oligotrophic medium [41], and marine agar [42]. The isolation of slow-growing actinomycetes needed all these different media to be supplemented with nalidixic acid (25 μg/mL), nystatin (25 μg/mL), and 0.2 μM pore size filtered cycloheximide (100 μg/mL). Nystatin and cycloheximide prevent fungal growth, while nalidixic acid prevents many fast-growing Gram-negative bacteria [43]. All media consisted of Difco Bacto agar (18 g/L) and were prepared in 1 L artificial sterile sea water [44]. *Micromonospora* sp. UR56 and *Actinokinospora* sp. EG49 were cultivated on ISP2 medium. The inoculated agar plates were incubated at 30 °C for a long time, ranging from 6 to 8 weeks. Distinct colony morphotypes were selected and re-streaked many times until free of any contaminants.

3.4. Molecular Identification

16S rRNA gene amplification, cloning, and sequencing were performed using the universal primers 27F and 1492R according to Hentschel et al. [45]. Chimeric sequences were detected using the Pintail program [46]. The genus-level affiliation of the sequence was confirmed using the Ribosomal Database Project Classifier. The genus-level identification of all the sequences was performed using RDP Classifier (-g 16srrna, -f allrank) and confirmed with the SILVA Incremental Aligner (SINA) (search and classify option) [47]. An alignment was evaluated again using the SINA web aligner (variability profile: bacteria). Gap-only positions were removed with trimAL (-noallgaps). For phylogenetic tree construction, the best fitting model was first estimated using Model Generator. Visualization was performed using Interactive Tree of Life.

3.5. Microbial Fermentation and Extract Preparation

Micromonospora sp. UR 56 and *Actinokineospora* sp. EG49 were isolated from Red Sea sponges *Callyspongia* sp and *Spheciospongia vagabunda*, respectively. Each microbial strain was fermented in 10 Erlenmeyer flasks (2 L), each containing 1 L of ISP2 medium and incubated at 30 °C with shaking (150 rpm) for 14 days. For the co-fermentation experiment, 10 mL of 5 day old culture of *Micromonospora* sp. UR56 was transferred into 10 Erlenmeyer flasks (2 L), each containing 1 L of ISP2 medium inoculated with 10 mL of 5 day old culture of *Actinokineospora* sp. EG49. After fermentation of axenic cultures and co-culture, filtration was performed, and the supernatant was extracted with ethyl acetate (1.5 L) to give the ethyl acetate soluble fraction (700 mg).

3.6. LC-HR/MS Metabolomic Analysis

LC–HR–ESI–MS metabolomics analyses were performed as previously described by Abdelmohsen et al. [48]. Ethyl acetate soluble fraction 1 mg/mL in MeOH was uploaded and analyzed using an Accela HPLC (Thermo Fisher Scientific, Karlsruhe, Germany) combined with UV–visible detector and Exactive-Orbitrap mass spectrometer (Thermo Fisher Scientific, Karlsruhe, Germany) using an HPLC column (an ACE C18, 75 mm × 3.0 mm, 5 μM column (Hichrom Limited, Reading, UK). The gradient elution was carried out at 300 μL/min for 30 min using purified water (A) and acetonitrile (B) with 0.1% formic acid in each mobile phase. The gradient program started with 10% B, increased gradually to 100% B, and continued isocratic for 5 min before linearly decreasing back to 10% B for 1 min. The total analysis period for each fraction was 45 min. The injection volume was 10 μL and the column temperature was maintained at 20 °C. High-resolution mass spectrometry was performed in both negative and positive ionization modes with a spray voltage of 4.5 kV and capillary temperature of 320 °C. The mass range was maintained at 150–1500 *m/z*. All positive and negative ionization files used to cover the highest number of metabolites were subjected to data mining software MZmine 2.10 (Okinawa Institute of Science and Technology Graduate University, Japan) for deconvolution, peak picking, alignment, deisotoping, and molecular formula prediction. Dictionary of natural products (DNP), Marinlit, and METLIN databases were used for identification of all metabolites.

3.7. Isolation and Purification of Induced Metabolites

The crude ethyl acetate (EtOAc) soluble fraction (700 mg, obtained from co-fermentation) was chromatographed on a Sephadex LH20 (32-64 μM, 100 × 25 mm) column using MeOH/H_2O (80 : 20%) to afford five main fractions (Fr.1 - Fr.5). Fr.2 (200 mg) was chromatographed over a silica gel column (CC). Gradient elution was performed using a gradient mixture of DCM : EtOAc (100 : 0 to 0 : 100) followed by 100% MeOH to afford 13 subfractions. SubFr.5 (30 mg) was further chromatographed on a Sephadex LH20 using Hex: DCM (50:50) as mobile phase to yield compound **3** (15 mg, 2.14% crude weight). SubFr.7 was washed several times with DCM, resulting in crystallization and affording compound **9** (10 mg, 1.42% crude weight). Fr.3 (175 mg) was chromatographed over a silica gel column with *n*-hexane: EtOAc (100:0 to 0:100) to yield nine subfractions (Fr.3-1 to Fr.3-9). Fr.3-7 and Fr.3-8 (30 mg and 20 mg) were further chromatographed over Sephadex LH20 using Hex: CH_2CL_2 (50:50) as mobile phase to afford pure compounds **1** and **2** (12 mg and 15 mg, 1.71 and 2.14% crude weight), respectively. Fr.5 (75 mg) was further chromatographed on a Sephadex LH20 using MeOH as mobile phase to afford pure compound **10** (8 mg, 1.14% crude weight).

3.8. Assessment of Antibacterial Activity

The antibacterial activity of all isolated pure compounds (**1–3**, **9**, and **10**) was evaluated on Gram-positive pathogenic bacteria such as *Bacillus subtilis* (ATCC29212) and *Staphylococcus aureus* (ATCC9144 (and Gram-negative pathogenic bacteria such as *Escherichia coli.* (ATCC25922) and *Pseudomonas aeruginosa* (ATCC27853), as previously reported [49]. The tests were carried out in 96 well flat polystyrene plates. First, 10 μL of each isolated compound (15 μM) was transferred to 80 μL of lysogeny broth followed by the addition of 10 μL of bacterial suspension at log phase, and then all inoculated plates were incubated overnight at 37 °C. After incubation, the positive effect of the tested compounds was detected as clearance in the wells, and in compounds that showed no activity on the bacteria, the growth medium in the wells seemed opaque. The absorbance was observed after 20 h at OD600 in a Spectrostar Nano Microplate Reader (BMG LABTECH GmbH, Allmendgrun, Germany). The positive control was pathogenic bacteria plus distilled water, while the negative control was growth medium plus distilled water.

3.9. Assessment of Antibiofilm Activity

The biofilm-inhibitory activities of the isolated pure compounds (**1–3**, **9**, and **10**) were measured using 96 well flat polystyrene plates against four clinical microbes comprising Gram-positive (*Bacillus subtilis* and *Staphylococcus aureus*) and Gram-negative (*Escherichia coli* and *Pseudomonas aeruginosa*) pathogenic bacteria, according to Antunes, et al. [50] with some modifications. Briefly, each well was filled with 180 µL lysogeny broth (LB broth) and then inoculated with 10 µL of pathogenic bacteria followed by addition of 10 µL of sample (2 µM) along with control (without test sample). All inoculated plates were then incubated overnight at 37 °C and, after incubation, content in the wells was removed and wells were rinsed with 200 µL of phosphate-buffered saline (PBS) pH 7.2 to eliminate free-floating microbes and left to dry under sterilized laminar flow for 1 h. For staining, 200 µL/well of crystal violet (0.1%, w/v) was added, left for 1 h, and then excessive stain was removed and plates retained for drying. Further, dried plates were rinsed with 95% ethanol and optical density was measured at 570 nm using a Spectrostar Nano Microplate Reader (BMG LABTECH GmbH, Allmendgrun, Germany).

3.10. Assessment of Cytotoxic Activity

MTT Assay

This assay was performed using different human cancer cell lines, including mammary gland breast cancer (MCF-7), colorectal carcinoma (HCT-116), human lung cancer cell line (WI38), and hepatocellular carcinoma (HePG-2). The cell lines were purchased from ATCC by (VACSERA), Cairo, Egypt. Doxorubicin was used as a positive control drug. The cell lines were used to detect the inhibitory effects of isolated compounds on cell growth. Cell viability was determined via colorimetric assay (MTT), which is based on the transformation of the yellow tetrazolium bromide to a purple formazan by mitochondrial succinate dehydrogenase. Cell lines were cultivated in rpmI-1640 medium supplemented with 10% fetal bovine serum, 100 units/mL penicillin, and 100 µg/mL streptomycin incubated at 37 °C under 5% CO_2. The cell lines were plated in a 96 well microtiter plate at a concentration of 1.0×10^4 cells/well and incubated at 37 °C under 5% CO_2 for 2 days. Subsequently, the cells were mixed with different concentrations of the isolated compounds and incubated for a further 24 h, and then treated with 20 µL of MTT solution at a concentration of 5 mg/mL and incubated for 4 h. The purple formazan precipitates were dissolved in 100 µL of dimethyl sulfoxide (DMSO) and the optical density for each well was measured and recorded at an absorbance of 570 nm using a plate reader (EXL 800, Biotek®, Winooski, Vermont, USA). The relative cell viability as a percentage was calculated as A570 of treated samples/A570 of untreated sample × 100.

3.11. Enzyme Assays

DNA gyrase (type II topoisomerase) and topoisomerase IV (ParE) add negative supercoils into DNA using ATP hydrolysis as a source of energy. They are considered crucial bacterial enzymes that are absent in eukaryotes. DNA-gyrase-B- and ParE-inhibitory activities were evaluated using the Inspiralis assay kit (Inspiralis®, London, UK) on streptavidin-coated 96 well microtiter plates (Thermo Scientific, Hamburg, Germany), according to the manufacturer's protocol [51]. The assay detects the ability of the isolated compounds to prevent the ATPase activity of both gyrase-B and ParE subunits. On the other hand, pyruvate kinase (PK) has been discovered to be an important hub protein in the interactome of MRSA [52]. The PK-inhibitory activity of the isolated compounds was assessed according to a previously reported method [53].

3.12. Docking Analysis

Staphylococcus Gyr-B and PK crystal structures with the of PDB codes 3g7b and 3T0T, respectively, were used. Docking experiments were performed using AutoDock Vina docking software [54]. Such docking engines deal with the receptor as a rigid structure and the ligand as a flexible structure during their calculations. The co-crystallized ligands were utilized to assign the binding sites. The ligand-to-binding-site shape-matching root mean square threshold was set to 2.0 Å. The interaction energies were determined using the Charmm force field (CFF) (v.1.02) with 10.0 Å as a non-bonded cutoff distance and distance-dependent dielectric. Subsequently, 5.0 Å was set as an energy grid extending from the binding site. The tested compounds were energy-minimized inside the selected binding pocket. The editing and visualization of the generated binding poses were performed using Pymol 2.3 software (Schrödinger, München, Germany).

3.13. Structuraal Elucidation of Isolated Compounds 1–3, 9, and 10.

The structures of the known induced metabolites were confirmed by comparison of their spectroscopic (HRMS and ^{1}H, ^{13}C and 2D-NMR) data with the published literature data as **1**, dimethylphenazine-1,6-dicarboxylate [55]; **2**, phencomycin [55–57]; **3**, tubermycinB [57–60]; **9**, N-(2-hydroxyphenyl)-acetamide [24,56]; and **10**, *p*-anisamide [61].

4. Conclusions

Actinomycetes genomes, in particular the order Actinomycetales, consist of numerous biosynthetic gene clusters encoding for secondary metabolites that have not yet been detected under standard laboratory conditions, and that need to be cultivated using elicitation approaches such as co-cultivation. In the present work, two Red-Sea-sponge-associated actinomycetes, *Micromonospora* sp. UR56 and *Actinokineospora* sp. EG49, were subjected to co-cultivation. Metabolomic profiling of both axenic actinomycetes and their co-culture revealed that the latter was more diverse in its induced metabolites profile, particularly in terms of phenazine derivatives. After isolation of the major induced metabolites (**1–3**, **9**, and **10**), their possible antibacterial, antibiofilm, and cytotoxic activities were assessed. Phenazine derivatives **1–3** were significantly active as antibacterial and antibiofilm agents with mild cytotoxicity. Subsequently, they were tested against Gyr-B and PK as possible molecular targets. The presence of carboxylic acid or an ester moiety at C-1 and C-6 was found to be crucial for the activity spectra of these compounds, where the antibacterial and antibiofilm effects were flipped from Gram-positive to Gram-negative strains, and their enzyme-inhibitory activities were significantly decreased upon removal of either carboxylic acid or the ester moiety at C-6. These findings could represent a good starting point from which to develop further phenazine-based antibiotics. Additionally, they highlight microbial mixed fermentation as a simple and effective approach to produce new antimicrobial agents through the induction of otherwise cryptic bacterial biogenetic pathways.

Supplementary Materials: The following are available online at http://www.mdpi.com/1660-3397/18/5/243/s1, Tables S1–S3: The dereplication results of the ethyl acetate fraction of micromonospora UR 56, Figures S1-S29: 1D and 2D NMR spectra of isolated compounds, Figures S30–S32: Dereplicated metabolites from metabolomic analysis.

Author Contributions: Conceptualization, M.S.H., R.M., H.M.H., A.M.S., M.E.R. and M.M.F.; methodology, M.M.F. and A.M.S.; data curation, A.A.H., H.M.H., U.R.A., M.E.R., A.M.S., H.A.A. and S.F.A.; original draft preparation, M.M.F.; writing, review and editing, all authors. All authors have read and agreed to the published version of the manuscript.

Funding: This research received no external funding.

Conflicts of Interest: The authors declare there is no conflict of interest.

References

1. Pettit, R.K. Mixed fermentation for natural product drug discovery. *Appl. Microbiol. Biotechnol.* **2009**, *83*, 19–25. [CrossRef] [PubMed]

2. Engelhardt, K.; Degnes, K.F.; Kemmler, M.; Bredholt, H.; Fjærvik, E.; Klinkenberg, G.; Sletta, H.; Ellingsen, T.E.; Zotchev, S.B. Production of a new thiopeptide antibiotic, TP-1161, by a marine *Nocardiopsis* species. *Appl. Environ. Microbiol.* **2010**, *76*, 4969–4976. [CrossRef] [PubMed]

3. Olano, C.; Méndez, C.; Salas, J.A. Antitumor compounds from marine actinomycetes. *Mar. Drugs* **2009**, *7*, 210–248. [CrossRef] [PubMed]

4. Pimentel-Elardo, S.M.; Kozytska, S.; Bugni, T.S.; Ireland, C.M.; Moll, H.; Hentschel, U. Anti-parasitic compounds from *Streptomyces* sp. strains isolated from Mediterranean sponges. *Mar. Drugs* **2010**, *8*, 373–380. [CrossRef]

5. Abdelmohsen, U.R.; Szesny, M.; Othman, E.M.; Schirmeister, T.; Grond, S.; Stopper, H.; Hentschel, U. Antioxidant and anti-protease activities of diazepinomicin from the sponge-associated *Micromonospora* strain RV115. *Mar. Drugs* **2012**, *10*, 2208–2221. [CrossRef]

6. Wolfender, J.-L.; Marti, G.; Ferreira Queiroz, E. Advances in techniques for profiling crude extracts and for the rapid identificationof natural products: Dereplication, quality control and metabolomics. *Curr. Org. Chem.* **2010**, *14*, 1808–1832. [CrossRef]

7. Wang, J.; Lin, W.; Wray, V.; Lai, D.; Proksch, P. Induced production of depsipeptides by co-culturing *Fusarium tricinctum* and *Fusarium begoniae*. *Tetrahedron Lett.* **2013**, *54*, 2492–2496. [CrossRef]

8. Nett, M.; Ikeda, H.; Moore, B.S. Genomic basis for natural product biosynthetic diversity in the actinomycetes. *Nat. Prod. Rep.* **2009**, *26*, 1362–1384. [CrossRef]

9. Winter, J.M.; Behnken, S.; Hertweck, C. Genomics-inspired discovery of natural products. *Curr. Opin. Chem. Biol.* **2011**, *15*, 22–31. [CrossRef]

10. Oh, D.-C.; Jensen, P.R.; Kauffman, C.A.; Fenical, W. Libertellenones A–D: Induction of cytotoxic diterpenoid biosynthesis by marine microbial competition. *Bioorg. Med. Chem.* **2005**, *13*, 5267–5273. [CrossRef]

11. Scherlach, K.; Hertweck, C. Triggering cryptic natural product biosynthesis in microorganisms. *Org. Biomol. Chem.* **2009**, *7*, 1753–1760. [CrossRef] [PubMed]

12. Schroeckh, V.; Scherlach, K.; Nützmann, H.-W.; Shelest, E.; Schmidt-Heck, W.; Schuemann, J.; Martin, K.; Hertweck, C.; Brakhage, A.A. Intimate bacterial–fungal interaction triggers biosynthesis of archetypal polyketides in *Aspergillus nidulans*. *Proc. Natl. Acad. Sci. USA* **2009**, *106*, 14558–14563. [CrossRef] [PubMed]

13. Chiang, Y.-M.; Chang, S.-L.; Oakley, B.R.; Wang, C.C.C. Recent advances in awakening silent biosynthetic gene clusters and linking orphan clusters to natural products in microorganisms. *Curr. Opin. Chem. Biol.* **2011**, *15*, 137–143. [CrossRef] [PubMed]

14. El-Hawary, S.S.; Sayed, A.M.; Mohammed, R.; Khanfar, M.; Rateb, M.E.; Mohammed, T.A.; Hajjar, D.; Hassan, H.M.; Gulder, T.A.M.; Abdelmohsen, U.R. New Pim-1 kinase inhibitor from the co-culture of two sponge-associated actinomycetes. *Front. Chem.* **2018**, *6*, 538. [CrossRef]

15. Cueto, M.; Jensen, P.R.; Kauffman, C.; Fenical, W.; Lobkovsky, E.; Clardy, J. Pestalone, a new antibiotic produced by a marine fungus in response to bacterial challenge. *J. Nat. Prod.* **2001**, *64*, 1444–1446. [CrossRef]

16. Thissera, B.; Alhadrami, H.A.; Hassan, M.H.A.; Hassan, H.M.; Bawazeer, M.; Yaseen, M.; Belbahri, L.; Rateb, M.E.; Behery, F.A. Induction of cryptic antifungal pulicatin derivatives from *Pantoea agglomerans* by microbial co-culture. *Biomolecules* **2020**, *10*, 268. [CrossRef]

17. Oh, D.-C.; Kauffman, C.A.; Jensen, P.R.; Fenical, W. Induced production of emericellamides A and B from the marine-derived fungus *Emericella* sp. In competing co-culture. *J. Nat. Prod.* **2007**, *70*, 515–520. [CrossRef]

18. McDonald, M.; Mavrodi, D.V.; Thomashow, L.S.; Floss, H.G. Phenazine biosynthesis in *Pseudomonas fluorescens*: Branchpoint from the primary shikimate biosynthetic pathway and role of phenazine-1, 6-dicarboxylic Acid. *J. Am. Chem. Soc.* **2001**, *123*, 9459–9460. [CrossRef]

19. Liu, B.; Liu, K.; Lu, Y.; Zhang, D.; Yang, T.; Li, X.; Ma, C.; Zheng, M.; Wang, B.; Zhang, G. Systematic evaluation of structure-activity relationships of the riminophenazine class and discovery of a C2 pyridylamino series for the treatment of multidrug-resistant tuberculosis. *Molecules* **2012**, *17*, 4545–4559. [CrossRef]

20. Kunz, A.; Labes, A.; Wiese, J.; Bruhn, T.; Bringmann, G.; Imhoff, J. Nature's lab for derivatization: New and revised structures of a variety of streptophenazines produced by a sponge-derived *Streptomyces* strain. *Mar. Drugs* **2014**, *12*, 1699–1714. [CrossRef]

21. Reddy, V.M.; O'Sullivan, J.F.; Gangadharam, P.R.J. Antimycobacterial activities of riminophenazines. *J. Antimicrob. Chemother.* **1999**, *43*, 615–623. [CrossRef] [PubMed]

22. Mavrodi, D.V.; Blankenfeldt, W.; Thomashow, L.S. Phenazine compounds in fluorescent *Pseudomonas* spp. biosynthesis and regulation. *Annu. Rev. Phytopathol.* **2006**, *44*, 417–445. [CrossRef] [PubMed]

23. Spicer, J.A.; Gamage, S.A.; Rewcastle, G.W.; Finlay, G.J.; Bridewell, D.J.A.; Baguley, B.C.; Denny, W.A. Bis (phenazine-1-carboxamides): Structure– activity relationships for a new class of dual topoisomerase I/II-directed anticancer drugs. *J. Med. Chem.* **2000**, *43*, 1350–1358. [CrossRef] [PubMed]

24. Dashti, Y.; Grkovic, T.; Abdelmohsen, U.R.; Hentschel, U.; Quinn, R.J. Production of induced secondary metabolites by a co-culture of sponge-associated actinomycetes, *Actinokineospora* sp. EG49 and *Nocardiopsis* sp. RV163. *Mar. Drugs* **2014**, *12*, 3046–3059. [CrossRef]

25. Boumehira, A.Z.; El-Enshasy, H.A.; Hacène, H.; Elsayed, E.A.; Aziz, R.; Park, E.Y. Recent progress on the development of antibiotics from the genus *Micromonospora*. *Biotechnol. bioprocess Eng.* **2016**, *21*, 199–223. [CrossRef]

26. Kuncharoen, N.; Kudo, T.; Ohkuma, M.; Tanasupawat, S. *Micromonospora azadirachtae* sp. nov., isolated from roots of Azadirachta indica A. Juss. var. siamensis Valeton. *Antonie van Leeuwenhoek, Int. J. Gen. Mol. Microbiol.* **2019**, *112*, 253–262. [CrossRef]

27. Borrero, N.V.; Bai, F.; Perez, C.; Duong, B.Q.; Rocca, J.R.; Jin, S.; Huigens, R.W., III. Phenazine antibiotic inspired discovery of potent bromophenazine antibacterial agents against *Staphylococcus aureus* and *Staphylococcus epidermidis*. *Org. Biomol. Chem.* **2014**, *12*, 881–886. [CrossRef]

28. Zhuo, S.-T.; Li, C.-Y.; Hu, M.-H.; Chen, S.-B.; Yao, P.-F.; Huang, S.-L.; Ou, T.-M.; Tan, J.-H.; An, L.-K.; Li, D. Synthesis and biological evaluation of benzo [a] phenazine derivatives as a dual inhibitor of topoisomerase I and II. *Org. Biomol. Chem.* **2013**, *11*, 3989–4005. [CrossRef]

29. Vicker, N.; Burgess, L.; Chuckowree, I.S.; Dodd, R.; Folkes, A.J.; Hardick, D.J.; Hancox, T.C.; Miller, W.; Milton, J.; Sohal, S. Novel angular benzophenazines: Dual topoisomerase I and topoisomerase II inhibitors as potential anticancer agents. *J. Med. Chem.* **2002**, *45*, 721–739. [CrossRef]

30. Adjei, A.A.; Charron, M.; Rowinsky, E.K.; Svingen, P.A.; Miller, J.; Reid, J.M.; Sebolt-Leopold, J.; Ames, M.M.; Kaufmann, S.H. Effect of pyrazoloacridine (NSC 366140) on DNA topoisomerases I and II. *Clin. cancer Res.* **1998**, *4*, 683–691.

31. Yi, L.; Lü, X. New strategy on antimicrobial-resistance: Inhibitors of DNA replication enzymes. *Curr. Med. Chem.* **2019**, *26*, 1761–1787. [CrossRef] [PubMed]

32. Zoraghi, R.; Worrall, L.; See, R.H.; Strangman, W.; Popplewell, W.L.; Gong, H. Methicillin-resistant Staphylococcus aureus (MRSA) pyruvate kinase as a target for bis-indole alkaloids with antibacterial activities. *J. Biol. Chem.* **2011**, *286*, 44716–44725. [CrossRef] [PubMed]

33. Ohlsen, K.; Donat, S. The impact of serine/threonine phosphorylation in *Staphylococcus aureus*. *Int. J. Med. Microbiol.* **2010**, *300*, 137–141. [CrossRef] [PubMed]

34. Vasu, D.; Sunitha, M.M.; Srikanth, L.; Swarupa, V.; Prasad, U.V.; Sireesha, K.; Yeswanth, S.; Kumar, P.S.; Venkatesh, K.; Chaudhary, A. In *Staphylococcus aureus* the regulation of pyruvate kinase activity by serine/threonine protein kinase favors biofilm formation. *3 Biotech* **2015**, *5*, 505–512. [CrossRef] [PubMed]

35. Wang, J.; Jiao, H.; Meng, J.; Qiao, M.; Du, H.; He, M.; ming, K.; Liu, J.; Wang, D.; Wu, Y. Baicalin inhibits biofilm formation and the quorum-sensing system by regulating the MsrA drug efflux pump in *Staphylococcus saprophyticus*. *Front. Microbiol.* **2019**, *10*, 2800. [CrossRef]

36. Benedetto Tiz, D.; Kikelj, D.; Zidar, N. Overcoming problems of poor drug penetration into bacteria: Challenges and strategies for medicinal chemists. *Expert Opin. Drug Discov.* **2018**, *13*, 497–507. [CrossRef]

37. Ronkin, S.M.; Badia, M.; Bellon, S.; Grillot, A.-L.; Gross, C.H.; Grossman, T.H.; Mani, N.; Parsons, J.D.; Stamos, D.; Trudeau, M. Discovery of pyrazolthiazoles as novel and potent inhibitors of bacterial gyrase. *Bioorg. Med. Chem. Lett.* **2010**, *20*, 2828–2831. [CrossRef]

38. Axerio-Cilies, P.; See, R.H.; Zoraghi, R.; Worral, L.; Lian, T.; Stoynov, N.; Jiang, J.; Kaur, S.; Jackson, L.; Gong, H. Cheminformatics-driven discovery of selective, nanomolar inhibitors for staphylococcal pyruvate kinase. *ACS Chem. Biol.* **2011**, *7*, 350–359. [CrossRef]

39. Mincer, T.J.; Jensen, P.R.; Kauffman, C.A.; Fenical, W. Widespread and persistent populations of a major new marine actinomycete taxon in ocean sediments. *Appl. Environ. Microbiol.* **2002**, *68*, 5005–5011. [CrossRef]

40. Shirling, E.B.T.; Gottlieb, D. Methods for characterization of *Streptomyces* species. *Int. J. Syst. Bacteriol.* **1966**, *16*, 313–340. [CrossRef]

41. Olson, J.B.; Lord, C.C.; McCarthy, P.J. Improved recoverability of microbial colonies from marine sponge samples. *Microb. Ecol.* **2000**, *40*, 139–147. [CrossRef] [PubMed]

42. Weiner, R.M.; Segall, A.M.; Colwell, R.R. Characterization of a marine bacterium associated with *Crassostrea virginica* (the eastern oyster). *Appl. Environ. Microbiol.* **1985**, *49*, 83–90. [CrossRef] [PubMed]

43. Webster, N.S.; Wilson, K.J.; Blackall, L.L.; Hill, R.T. Phylogenetic diversity of bacteria associated with the marine sponge *Rhopaloeides odorabile*. *Appl. Environ. Microbiol.* **2001**, *67*, 434–444. [CrossRef] [PubMed]

44. Lyman, J.; Fleming, R.H. Composition of sea water. *J. mar. Res* **1940**, *3*, 134–146.

45. Hentschel, U.; Hopke, J.; Horn, M.; Friedrich, A.B.; Wagner, M.; Hacker, J.; Moore, B.S. Molecular evidence for a uniform microbial community in sponges from different oceans. *Appl. Environ. Microbiol.* **2002**, *68*, 4431–4440. [CrossRef] [PubMed]

46. Ashelford, K.E.; Chuzhanova, N.A.; Fry, J.C.; Jones, A.J.; Weightman, A.J. At least 1 in 20 16S rRNA sequence records currently held in public repositories is estimated to contain substantial anomalies. *Appl. Environ. Microbiol.* **2005**, *71*, 7724–7736. [CrossRef]

47. Pruesse, E.; Peplies, J.; Glöckner, F.O. SINA: Accurate high-throughput multiple sequence alignment of ribosomal RNA genes. *Bioinformatics* **2012**, *28*, 1823–1829. [CrossRef]

48. Abdelmohsen, U.; Cheng, C.; Viegelmann, C.; Zhang, T.; Grkovic, T.; Ahmed, S.; Quinn, R.; Hentschel, U.; Edrada-Ebel, R. Dereplication strategies for targeted isolation of new antitrypanosomal actinosporins A and B from a marine sponge associated-*Actinokineospora* sp. EG49. *Mar. Drugs* **2014**, *12*, 1220–1244. [CrossRef]

49. Ingebrigtsen, R.A.; Hansen, E.; Andersen, J.H.; Eilertsen, H.C. Light and temperature effects on bioactivity in diatoms. *J. Appl. Phycol.* **2016**, *28*, 939–950. [CrossRef]

50. Antunes, A.L.S.; Trentin, D.S.; Bonfanti, J.W.; PINTO, C.C.F.; PEREZ, L.R.R.; Macedo, A.J.; Barth, A.L. Application of a feasible method for determination of biofilm antimicrobial susceptibility in *staphylococci*. *Apmis* **2010**, *118*, 873–877. [CrossRef]

51. Durcik, M.; Tammela, P.; Barančoková, M.; Tomašič, T.; Ilaš, J.; Kikelj, D.; Zidar, N. Synthesis and evaluation of N-phenylpyrrolamides as DNA gyrase B Inhibitors. *ChemMedChem* **2018**, *13*, 186–198. [CrossRef] [PubMed]

52. Suzuki, K.; Ito, S.; Shimizu-Ibuka, A.; Sakai, H. Crystal structure of pyruvate kinase from Geobacillus stearothermophilus. *J. Biochem.* **2008**, *144*, 305–312. [CrossRef] [PubMed]

53. El-Sayed, M.T.; Zoraghi, R.; Reiner, N.; Suzen, S.; Ohlsen, K.; Lalk, M.; Altanlar, N.; Hilgeroth, A. Novel inhibitors of the methicillin-resistant *Staphylococcus aureus* (MRSA)-pyruvate kinase. *J. Enzyme Inhib. Med. Chem.* **2016**, *31*, 1666–1671. [CrossRef] [PubMed]

54. Sayed, A.M.; Alhadrami, H.A.; El-Hawary, S.S.; Mohammed, R.; Hassan, H.M.; Rateb, M.E.; Abdelmohsen, U.R.; Bakeer, W. Discovery of two brominated oxindole alkaloids as *Staphylococcal* DNA gyrase and pyruvate kinase inhibitors via inverse virtual screening. *Microorganisms* **2020**, *8*, 293. [CrossRef] [PubMed]

55. Chatterjee, S.; Vijayakumar, E.K.S.; Franco, C.M.M.; Maurya, R.; Blumbach, J.; Ganguli, B.N. Phencomycin, a new antibiotic from a *Streptomyces* species HIL Y-9031725. *J. Antibiot. (Tokyo)*. **1995**, *48*, 1353–1354. [CrossRef]

56. Pusecker, K.; Laatsch, H.; Helmke, E.; Weyland, H. Dihydrophencomycin methyl ester, a new phenazine derivative from a marine *streptomycete*. *J. Antibiot. (Tokyo)*. **1997**, *50*, 479–483. [CrossRef]

57. Cheng, C.; Othman, E.M.; Fekete, A.; Krischke, M.; Stopper, H.; Edrada-Ebel, R.A.; Mueller, M.J.; Hentschel, U.; Abdelmohsen, U.R. Strepoxazine A, a new cytotoxic phenoxazin from the marine sponge-derived bacterium *Streptomyces* sp. SBT345. *Tetrahedron Lett.* **2016**, *57*, 4196–4199. Available online: http://dx.doi.org/10.1016/j.tetlet.2016.08.005. [CrossRef]

58. Geiger, A.; Keller-Schierlein, W.; Brandl, M.; Zahner, H. Metabolites of microorganisms. 247 phenazines from *Streptomyces antibioticus*, strain Tü 2706. *J. Antibiot.* **1988**, *41*, 1542–1551. [CrossRef]

59. Gebhardt, K.; Schimana, J.; Krastel, P.; Dettner, K.; Rheinheimer, J.; Zeeck, A.; Fiedler, H.-P. Endophenazines AD, new phenazine antibiotics from the Arthropod associated endosymbiont *Streptomyces anulatus*. *J. Antibiot.* **2002**, *55*, 794–800. [CrossRef]

60. Lee, H.-S.; Kang, J.S.; Choi, B.-K.; Lee, H.-S.; Lee, Y.-J.; Lee, J.; Shin, H.J. Phenazine derivatives with anti-inflammatory activity from the deep-sea sediment-derived yeast-like fungus *Cystobasidium laryngis* IV17-028. *Mar. Drugs.* **2019**, *17*, 482. [CrossRef]

61. National Center for Biotechnology Information.pub Chemdatabase. 4-Methoxy Benzamide. Available online: https://pubchem.ncbi.nlm.nih.gov/compound/4-Methoxybenzamide (accessed on 15 April 2020).

Article

Bioactive Steroids from the Red Sea Soft Coral *Sinularia polydactyla*

Mohamed A. Tammam [1,2], Lucie Rárová [3,*], Marie Kvasnicová [3], Gabriel Gonzalez [3], Ahmed M. Emam [2], Aldoushy Mahdy [4], Miroslav Strnad [3], Efstathia Ioannou [1] and Vassilios Roussis [1,*]

[1] Section of Pharmacognosy and Chemistry of Natural Products, Department of Pharmacy, National and Kapodistrian University of Athens, Panepistimiopolis Zografou, 15771 Athens, Greece; mtammam@pharm.uoa.gr (M.A.T.); eioannou@pharm.uoa.gr (E.I.)

[2] Department of Biochemistry, Faculty of Agriculture, Fayoum University, Fayoum 63514, Egypt; ame01@fayoum.edu.eg

[3] Laboratory of Growth Regulators, Institute of Experimental Botany, The Czech Academy of Sciences, Faculty of Science, Palacký University, Šlechtitelů 27, CZ-78371 Olomouc, Czech Republic; kvasnicova@ueb.cas.cz (M.K.); gonzalez.gabriel@seznam.cz (G.G.); miroslav.strnad@upol.cz (M.S.)

[4] Department of Zoology, Faculty of Science, Al-Azhar University (Assiut Branch), Assiut 71524, Egypt; aldoushy@azhar.edu.eg

* Correspondence: lucie.rarova@upol.cz (L.R.); roussis@pharm.uoa.gr (V.R.); Tel.: +420-5856-34698 (L.R.); +30-210-727-4592 (V.R.)

Received: 16 November 2020; Accepted: 8 December 2020; Published: 10 December 2020

Abstract: Six new (**1, 2, 6, 8, 13,** and **20**) and twenty previously isolated (**3–5, 7, 9–12, 14–19,** and **21–26**) steroids featuring thirteen different carbocycle motifs were isolated from the organic extract of the soft coral *Sinularia polydactyla* collected from the Hurghada reef in the Red Sea. The structures and the relative configurations of the isolated natural products have been determined based on extensive analysis of their NMR and MS data. The cytotoxic, anti-inflammatory, anti-angiogenic, and neuroprotective activity of compounds **3–7, 9–12, 14–20,** and **22–26**, as well as their effect on androgen receptor-regulated transcription was evaluated in vitro in human tumor and non-cancerous cells. Steroids **22** and **23** showed significant cytotoxicity in the low micromolar range against the HeLa and MCF7 cancer cell lines, while migration of endothelial cells was inhibited by compounds **11, 12, 22,** and **23** at 20 µM. The results of the androgen receptor (AR) reporter assay showed that compound **11** exhibited the strongest inhibition of AR at 10 µM, while it is noteworthy that steroids **10, 16,** and **20** displayed increased inhibition of AR with decreasing concentrations. Additionally, compounds **11** and **23** showed neuroprotective activity on neuron-like SH-SY5Y cells.

Keywords: *Sinularia polydactyla*; soft coral; steroids; cytotoxic; anti-inflammatory; neuroprotective; androgen receptor

---

1. Introduction

The Red Sea, one of the warmest and most saline marine habitats, is an extension of the Indian Ocean, located between the Arabian Peninsula and Africa. The entire coastal reef complex extends along a 2000-km shoreline and is characterized by a high degree of chemodiversity, including more than 200 soft and 300 hard coral species [1]. Despite the diverse marine life hosted in the Red Sea, marine organisms from this ecosystem have not been thoroughly studied in comparison to other extended coral habitats, such as those encountered in the Great Barrier Reef or the Caribbean Sea [2].

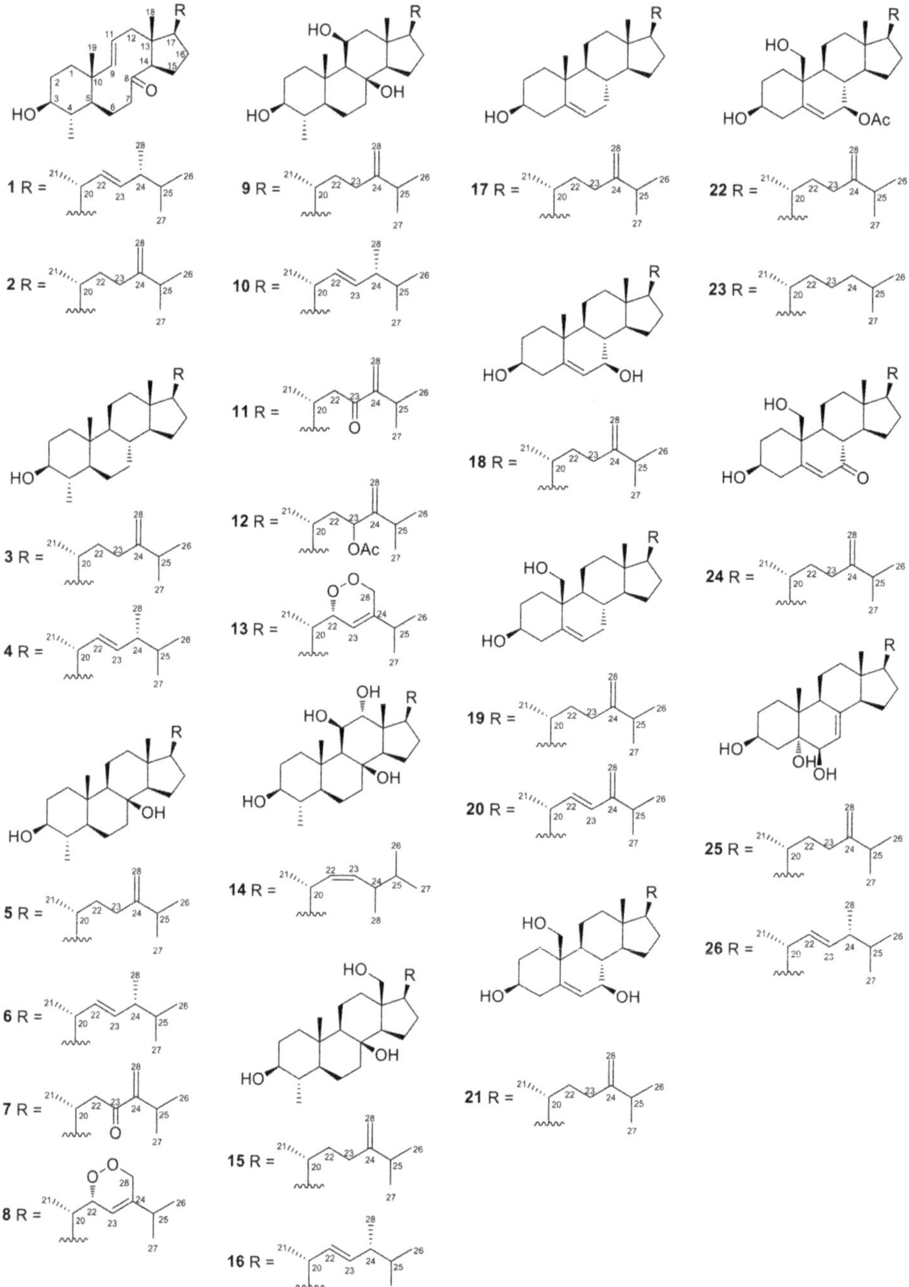

Figure 1. Chemical structures of compounds **1–26**.

Over the last 50 years, soft corals (Anthozoa, Gorgonacea) have been the subject of extensive chemical investigations which have resulted in the isolation of a large of number secondary metabolites, mainly sesquiterpenes, diterpenes, prostanoids, and highly functionalized steroids [3].

A significant number of these metabolites have exhibited potent biological properties, including cytotoxic, antibacterial, antifungal, anti-inflammatory, and antifouling activity [2]. Among soft corals, species of the genus *Sinularia* have been extensively studied as sources of new bioactive compounds, most often leading to the isolation of diterpenes and steroids with noteworthy levels of bioactivity [2,3].

In the context of our continuous interest for the isolation of bioactive metabolites from marine organisms, we recently had the opportunity to collect specimens of *Sinularia polydactyla* from the coastline of Hurghada in the Red Sea (Egypt) and investigate its chemical profile. Herein we report the isolation, structure elucidation, and evaluation of biological activity of six new (**1**, **2**, **6**, **8**, **13**, and **20**) and twenty previously isolated (**3–5**, **7**, **9–12**, **14–19**, and **21–26**) steroids (Figure 1).

2. Results and Discussion

2.1. Structure Elucidation of the Isolated Metabolites

A series of normal- and reversed-phase chromatographic separations of the organic extract of the soft coral *S. polydactyla* collected from the Egyptian Red Sea coastline at Hurghada allowed for the isolation of compounds **1–26**.

Compound **1**, isolated as a white amorphous solid, possessed the molecular formula $C_{29}H_{48}O_2$, as indicated by the HR-APCIMS and NMR data. The ^{13}C NMR and HSQC-DEPT spectra revealed the presence of 29 carbon atoms, corresponding to seven methyls, seven methylenes, twelve methines, and three non-protonated carbon atoms. Among them, evident were one carbonyl resonating at δ_C 213.1, four olefinic carbons resonating at δ_C 127.6, 132.6, 135.2, and 142.0 and an oxygenated carbon resonating at δ_C 76.6. In the ^1H NMR spectrum evident were two methyls on non-protonated carbons (δ_H 0.94 and 1.08), five methyls on tertiary carbons (δ_H 0.80, 0.81, 0.89, 0.99, and 1.04), one oxygenated methine (δ_H 3.09), and four olefinic methines (δ_H 5.15, 5.20, 5.28, and 5.41). Since the carbonyl moiety and the two carbon–carbon double bonds accounted for three of the six degrees of unsaturation, the molecular structure of **1** was determined as tricyclic. The spectroscopic features of metabolite **1** (Tables 1 and 2), in conjunction with the homonuclear and heteronuclear correlations observed in its HSQC, HMBC, and COSY spectra (Figure 2) suggested that compound **1** possessed a 8-oxo-3-hydroxy-4-methyl-8,9-seco steroidal nucleus with a $\Delta^{9,11}$ and a C_9H_{17} unsaturated side chain with a 1,2-disubstituted double bond between C-22 and C-23. Specifically, the correlations observed in the COSY spectrum identified three distinct spin systems, namely (a) a spin system starting from H-1 to H_2-7, incorporating H_3-29 which was coupled to H-4, (b) a short spin system from H-9 to H_2-12 through H-11, and (c) an extended branched spin system from H-14 to H_3-26, including H_3-21, H_3-27, and H_3-28 which were coupled to H-20, H-25, and H-24, respectively. The HMBC correlations of H_3-19 with C-1, C-5, C-9, and C-10 concluded the six-membered ring and positioned the first angular methyl on C-10, connecting at the same time spin systems (a) and (b). Additionally, the HMBC correlations of H_3-18 with C-12, C-13, C-14, and C-17 identified the five-membered ring and fixed the position of the second angular methyl on C-13, connecting spin systems (b) and (c). The HMBC correlations of H_2-6, H_2-7, and H-14 with the carbonyl carbon C-8 supported the cleavage of the C-8/C-9 bond, giving rise to the decalin ring and connecting spin systems (a) and (c). The geometry of the two double bonds was determined as E in both cases on the basis of the measured coupling constants ($J_{9,11}$ = 15.3 Hz and $J_{22,23}$ = 15.2 Hz). The enhancements of H-3/H-5, H-3/H_3-29, H-4/H_3-19, H-5/H-9, H-9/H-12α, H-9/H-14, H-11/H-12β, H-11/H_3-18, H-11/H_3-19, H-14/H-17, and H_3-18/H-20 observed in the NOESY spectrum (Figure 2) verified the *trans* fusion of the six-membered and the ten-membered rings, as well as the *trans* fusion of the latter with the five-membered ring and determined the relative configuration of the stereogenic centers. The R configuration at C-24 was proposed on the basis of the 0.3 ppm difference in the chemical shifts of C-26 and C-27 and the chemical shift of C-28 resonating at 17.6 ppm [4]. On the basis of the above, metabolite **1** was identified as (9*E*,22*E*,24*R*)-3β-hydroxy-4α,24-dimethyl-8,9-seco-5α-cholesta-9(11),22-dien-8-one.

Table 1. ^{1}H NMR data (δ in ppm, J in Hz) in $CDCl_3$ of compounds **1**, **2**, **4**, **6**, **8**, **13**, and **20**.

Position	1 [1]	2 [1]	4 [2]	6 [2]	8 [1]	13 [1]	20 [3]
1	1.33 m, 1.22 m	1.36 m, 1.24 m	1.71 m, 0.98 m	1.72 m, 0.93 m	1.72 m, 0.96 m	1.91 m, 1.00 m	1.91 m, 1.08 m
2	1.76 m, 1.48 m	1.76 m, 1.49 m	1.77 m, 1.46 m	1.77 m, 1.52 m	1.78 m, 1.52 m	1.79 m, 1.58 m	1.84 m, 1.40 m
3	3.09 td (10.1, 4.8)	3.10 m	3.06 td (10.5, 4.9)	3.04 td (10.2, 4.9)	3.06 td (10.5, 5.2)	3.05 td (10.8, 5.0)	3.56 m
4	1.27 m	1.28 m	1.27 m	1.31 m	1.33 m	1.41 m	2.37 m, 2.18 m
5	0.91 m	0.92 m	0.72 m	0.68 td (12.4, 2.9)	0.71 td (12.2, 2.2)	0.69 td (11.9, 2.2)	-
6	1.66 m, 1.60 m	1.66 m, 1.62 m	1.64 m, 1.47 m	1.50 m, 1.33 m	1.53 m, 1.33 m	1.56 m, 1.39 m	5.73 m
7	2.29 m, 1.75 m	2.48 m, 2.24 m	1.69 m, 1.50 m	1.63 m, 1.13 m	1.64 m, 1.19 m	1.71 m, 1.22 m	2.01 m, 1.51 m
8	-		1.28 m	-	-	-	1.82 m
9	5.28 d (15.3)	5.27 m	0.60 m	0.80 m	0.81 m	0.88 m	0.89 m
11	5.41 ddd (15.3, 11.1, 3.8)	5.41 m	1.47 m, 1.00 m	1.62 m, 1.49 m	1.62 m, 1.48 m	4.43 brd (1.9)	1.62 m, 1.53 m
12	2.47 m, 1.70 m	2.50 m, 1.72 m	1.92 m, 1.11 m	1.95 m, 1.16 m	1.93 m, 1.17 m	2.23 m, 1.37 m	2.01 m, 1.17 m
14	2.49 m	2.50 m	0.93 m	1.18 m	1.23 m	1.27 m	0.90 m
15	1.64 m, 1.48 m	1.64 m, 1.50 m	1.51 m, 1.02 m	1.45 m, 1.23 m	1.53 m, 1.33 m	1.60 m, 1.46 m	1.50 m, 1.05 m
16	1.70 m, 1.47 m	1.73 m, 1.49 m	1.63 m, 1.19 m	1.64 m, 1.23 m	1.99 m, 1.31 m	2.00 m, 1.35 m	1.64 m, 1.26 m
17	1.31 m	1.28 m	1.09 m	1.03 m	1.43 m	1.43 m	1.18 m
18	1.08 s	1.09 s	0.64 s	0.92 s	0.89 s	1.09 s	0.75 s
19	0.94 s	0.94 s	0.80 s	0.97 s	0.96 s	1.33 s	3.81 d (11.5), 3.59 d (11.5)
20	2.10 m	2.11 m	1.98 m	1.96 m	1.56 m	1.59 m	2.11 m
21	0.99 d (6.8)	0.93 d (6.1)	0.97 d (6.6)	0.94 d (6.4)	0.83 d (6.9)	0.85 d (7.0)	1.03 d (6.5)
22	5.15 dd (15.2, 8.1)	1.53 m, 1.14 m	5.12 dd (15.1, 7.6)	5.09 dd (15.2, 8.2)	4.72 brs	4.70 brs	5.56 dd (15.7, 8.7)
23	5.20 dd (15.2, 7.3)	2.08 m, 1.88 m	5.17 dd (15.1, 7.0)	5.16 dd (15.2, 7.4)	5.36 brs	5.36 brs	5.92 d (15.7)
24	1.83 m	-	1.83 m	1.80 m	-	-	-
25	1.44 m	2.21 m	1.44 m	1.42 m	2.24 septet (6.9)	2.24 septet (6.9)	2.53 septet (6.8)
26	0.81 d (6.8)	1.00 d (6.8)	0.81 d (6.8)	0.81 d (6.8)	1.05 d (6.9)	1.05 d (6.9)	1.06 d (6.8)
27	0.80 d (6.8)	1.01 d (6.9)	0.79 d (6.8)	0.79 d (6.8)	1.05 d (6.9)	1.06 d (6.9)	1.04 d (6.8)
28	0.89 d (6.8)	4.70 brs, 4.63 brs	0.88 d (6.8)	0.88 d (6.8)	4.62 d (15.7), 4.20 d (15.7)	4.59 d (15.7), 4.21 d (15.7)	4.83 brs, 4.79 brs
29	1.04 d (6.1)	1.04 d (6.1)	0.92 d (6.3)	0.95 d (6.8)	0.95 d (6.4)	0.96 d (6.4)	-

[1] Recorded at 600 MHz. [2] Recorded at 400 MHz. [3] Recorded at 950 MHz, - absence of value.

Table 2. ^{13}C NMR data (δ in ppm) in CDCl$_3$ of compounds **1**, **4**, **6**, **8**, **13**, and **20**.

Position	1 [1,2]	4 [3]	6 [1,4]	8 [1,2]	13 [2]	20 [1,5]
1	38.5	36.8	37.7	37.2	37.5	32.9
2	30.7	31.1	30.9	30.3	30.2	31.7
3	76.6	76.6	76.8	76.4	76.5	70.8
4	39.8	39.2	39.0	38.5	38.2	41.9
5	53.5	51.0	51.9	51.2	52.3	134.9
6	20.0	24.2	20.5	19.8	20.0	126.8
7	47.1	32.2	40.1	39.6	39.9	30.8
8	213.1	36.0	73.6	73.5	75.3	33.4
9	142.0	54.6	56.6	56.0	57.6	49.9
10	39.1	34.9	36.9	36.3	36.8	41.0
11	127.6	21.1	18.3	18.1	69.8	21.4
12	46.8	40.2	41.6	40.5	49.0	39.6
13	55.2	42.4	43.0	42.7	41.8	42.3
14	62.7	56.6	59.7	59.0	60.2	57.5
15	26.6	24.2	19.0	18.7	19.2	24.1
16	28.4	28.6	28.3	27.0	26.9	28.1
17	56.7	56.1	57.1	52.4	53.9	55.4
18	13.5	12.3	14.0	12.8	15.0	12.0
19	15.8	13.4	13.8	13.2	15.6	62.3
20	39.1	40.0	39.6	39.8	40.0	39.9
21	21.7	20.9	21.3	12.8	12.9	20.2
22	135.2	135.9	135.7	79.9	79.7	135.4
23	132.6	131.6	131.8	119.4	119.2	128.8
24	43.0	42.8	43.2	142.0	142.0	153.0
25	33.2	33.1	33.8	30.8	31.1	28.8
26	19.8	19.6	19.9	21.0	21.1	21.7
27	20.1	19.9	20.2	21.0	21.1	22.1
28	17.6	17.6	18.0	70.6	70.8	109.0
29	16.4	15.1	16.0	14.9	15.2	

[1] Chemical shifts were determined through HMBC correlations. [2] Recorded at 150 MHz. [3] Recorded at 50 MHz. [4] Recorded at 100 MHz. [5] Recorded at 237.5 MHz.

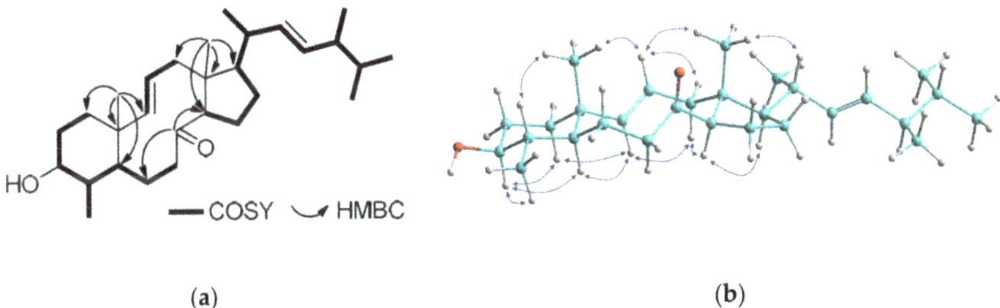

Figure 2. (a) COSY and key HMBC correlations and (b) key NOESY cross-peaks for compound **1**.

Compound **2**, isolated in minute amount as a white amorphous solid, displayed an ion peak at *m/z* 429.3725 (HR-APCIMS), corresponding to C$_{29}$H$_{49}$O$_2$ and consistent for [M + H]$^+$. Compound **2** shared quite similar spectroscopic features with **1**. In particular, all signals attributed to the steroidal nucleus, with the most prominent being the signals of the two angular methyls H$_3$-18 and H$_3$-19 (δ_H 1.09 and 0.94, respectively), the methyl at C-4 (δ_H 1.04), the oxymethine H-3 (δ_H 3.10), and the two olefinic protons H-9 and H-11 (δ_H 5.27 and 5.41, respectively), were also evident in the ^{1}H NMR spectrum of compound **2** (Table 1). The most significant difference observed was the replacement of the 1,2-disubstituted double bond in the side chain of **1** by a 1,1-disubstituted double bond (δ_H 4.63 and 4.70) in the side chain of **2**. The correlations observed in the COSY spectrum of **2** identified the relevant spin systems; however, metabolite **2** was proven unstable and degraded prior to the acquisition of heteronuclear NMR spectra. Nevertheless, the high structural similarity of **2** with compound **1** renders safe the proposed identification of **2** as (9*E*)-3*β*-hydroxy-4*α*,24-dimethyl-8,9-seco-5*α*-cholesta-9(11),24(28)-dien-8-one.

Compound **6**, isolated as a white amorphous solid, possessed the molecular formula $C_{29}H_{50}O_2$, as suggested by its HR-APCIMS and NMR data. The spectroscopic data of **6** were quite similar to those of the previously reported metabolites **5** and **7** (Tables 1 and 2). The presence of the two angular methyls at δ_H 0.92 and 0.97, the doublet methyl at δ_H 0.95, the oxygenated methine at δ_H 3.04 and the quaternary oxygenated carbon at δ_C 73.6, in conjunction with the correlations observed in the HMBC and COSY spectra (Figure 3a), verified the 3,8-dihydroxy-4-methyl steroidal nucleus. The side chain of compound **6** included four doublet methyls ($\delta_{H/C}$ 0.79/20.2, 0.81/19.9, 0.88/18.0, and 0.95/16.0) and two olefinic methines ($\delta_{H/C}$ 5.09/135.7 and 5.16/131.8) that was assigned on the basis of the COSY and HMBC correlations. The *E* geometry of the Δ^{22} double bond was supported by the large coupling constant of H-22/H-23 (*J* = 15.3 Hz). The enhancements of H-3/H-5, H-3/H$_3$-29, H-4/H$_3$-19, H-5/H-9, H-5/H$_3$-29, H-9/H-12α, H-9/H-14, H-12β/H$_3$-18, and H$_3$-18/H-20 observed in the NOESY spectrum verified the *trans* fusion of rings A/B, B/C, and C/D and suggested the axial orientation of the hydroxy group at C-8. The configuration at C-24 was proposed as *R* due to the fact that the difference in the chemical shifts of C-26 and C-27 was 0.3 ppm and that C-28 resonated at 18.0 ppm [4]. Therefore, compound **6** was identified as (22*E*,24*R*)-4*α*,24-dimethyl-5*α*-cholest-22-en-3*β*,8*β*-diol.

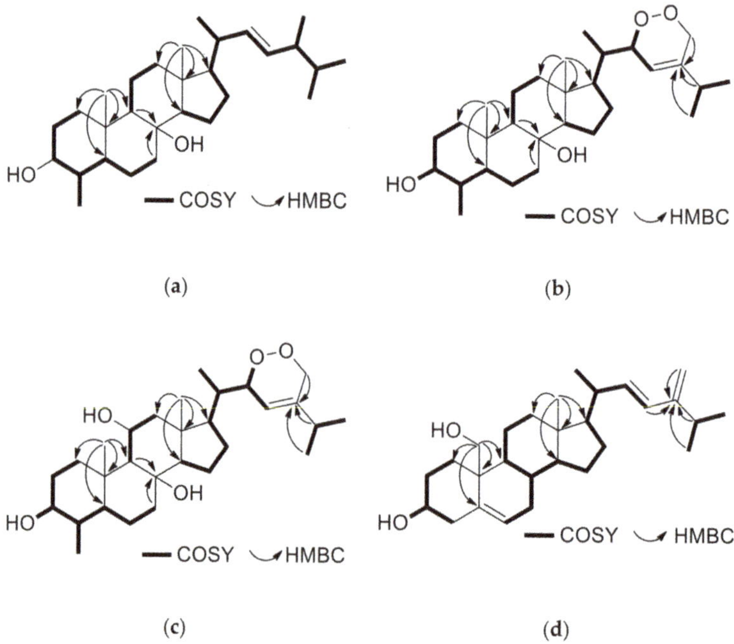

(a)

(b)

(c)

(d)

Figure 3. (**a**) COSY and important HMBC correlations for compound **6**. (**b**) COSY and important HMBC correlations for compound **8**. (**c**) COSY and important HMBC correlations for compound **13**. (**d**) COSY and important HMBC correlations for compound **20**.

Compound **8**, isolated as a white amorphous solid, exhibited an ion peak at *m/z* 459.3475 corresponding to $C_{29}H_{47}O_4$ and consistent with $[M - H]^-$. The high degree of similarity of the spectroscopic data of metabolite **8** (Tables 1 and 2) with those of **5–7** indicated the same 3,8-dihydroxy-4-methyl steroidal nucleus, further confirmed by the correlations observed in the HMBC and COSY spectra (Figure 3b). Taking into account that the steroidal nucleus of **8** accounts for four of the six degrees of unsaturation and the presence of one double bond on the side chain, the latter should also contain an additional ring. The ^{1}H and ^{13}C NMR signals corresponding to the side chain of compound **8** included three doublet methyls ($\delta_{H/C}$ 0.95/14.9, 1.05/21.0, and 1.05/21.0), one oxygenated methine ($\delta_{H/C}$ 4.72/79.9), one oxygenated methylene ($\delta_{H/C}$ 4.20, 4.62/70.6), one olefinic methine ($\delta_{H/C}$ 5.36/119.4), and one non-protonated olefinic carbon (δ_C 142.0). The COSY cross-peaks of H-20/H$_3$-21, H-20/H-22, H-22/H-23, H-25/H$_3$-26, and H-25/H$_3$-27, in combination with the HMBC correlations of H$_3$-21 with C-17, C-20, and C-22, of H-23, H-25, H$_3$-26, and H$_3$-27 with C-24 and of H$_2$-28 with C-23, C-24, and C-25 verified the side chain. In accordance with the literature, H-22 was assigned to be on the opposite side of H$_3$-21, as also suggested by the chemical shift of C-23 which resonated at 119.4 ppm. Instead, when H-22 and H$_3$-21 are co-planar, C-23 is shielded, resonating at 115–116 ppm [5]. Thus, metabolite **8** was identified as (23*E*)-22α,28-epidioxy-4α,24-dimethyl-5α-cholest-23-en-3β,8β-diol.

Compound **13** was isolated as a white amorphous solid. The ion peak at *m/z* 475.3424 observed in its HR-APCIMS was consistent with [M − H]$^-$, dictating the molecular formula C$_{29}$H$_{48}$O$_5$. The spectroscopic data of metabolite **13** related to the steroidal nucleus (Tables 1 and 2) displayed high similarity with those of **9–12**, suggesting a 3,8,11-trihydroxy-4-methyl steroidal nucleus that was further verified by the correlations observed in the COSY, HMBC and NOESY spectra (Figure 3c). Additionally, the NMR data concerning the side chain of compound **13** were rather similar to those of **8**, thus allowing for the identification of **13** as (23*E*)-22α,28-epidioxy-4α,24-dimethyl-5α-cholest-23-en-3β,8β,11β-triol.

Compound **20**, isolated as a white amorphous solid, had the molecular formula C$_{28}$H$_{44}$O$_2$, as indicated by its HR-ESIMS and NMR data. In the ^{1}H NMR spectrum of metabolite **20** evident were only one methyl on a non-protonated carbon (δ_H 0.75), three methyls on tertiary carbons (δ_H 1.03, 1.04 and 1.06), one hydroxymethylene (δ_H 3.59 and 3.81), one oxygenated methine (δ_H 3.56), three olefinic methines (δ_H 5.56, 5.73 and 5.92) and an exomethylene group (δ_H 4.79 and 4.83). The spectroscopic data of **20** (Tables 1 and 2) closely resembled those of the co-occurring **19**, with the main difference being the presence of an additional 1,2-disubstituted double bond in the side chain of **20**. The homonuclear and heteronuclear correlations observed in the COSY, HMBC, and NOESY spectra (Figure 3d) verified the 3,19-dihydroxy steroidal nucleus with a Δ^5 double bond and the proposed side chain, as well as the relative configuration of the stereogenic centers. The *E* geometry of the Δ^{22} double bond was assigned on the basis of the measured coupling constant between H-22 and H-23 (*J* = 15.7 Hz). On the basis of the above, metabolite **20** was identified as (22*E*)-24-methyl-cholesta-5,22,24(28)-trien-3β,19-diol.

Compounds **3**, **5**, **7**, **9–12**, **14–19**, and **21–26** were identified by comparison of their spectroscopic and physical characteristics with those reported in the literature as 4α,24-dimethyl-5α-cholest-24(28)-en-3β-ol (**3**) [6], (22*E*,24*R*)-4α,24-dimethyl-5α-cholest-22-en-3β-ol (**4**) [7], 4α,24-dimethyl-5α-cholest-24(28)-en-3β,8β-diol (**5**) [8], 23-oxo-4α,24-dimethyl-5α-cholest-24(28)-en-3β,8β-diol (**7**) [9], nebrosteroid M (4α,24-dimethyl-5α-cholest-24(28)-en-3β,8β,11β-triol, (**9**) [10], (22*E*,24*R*)-4α,24-dimethyl-5α-cholest-22-en-3β,8β,11β-triol (**10**) [4], nebrosteroid A (23-oxo-4α,24-dimethyl-5α-cholest-24(28)-en-3β,8β,11β-triol, (**11**) [11], 23ξ-acetoxy-4α,24-dimethyl-5α-cholest-24(28)-en-3β,8β,11β-triol (**12**) [4], (22*Z*)-4α,24ξ-dimethyl-5α-cholest-22-en-3β,8β, 11β,12α-tetraol (**14**) [4], 4α,24-dimethyl-5α-cholest-24(28)-en-3β, 8β, 18-triol (**15**) [4], (22*E*,24*R*)-4α,24-dimethyl-5α-cholest-22-en-3β,8β,18-triol (**16**) [4], 24-methyl-cholesta-5,24(28)-dien-3β-ol (**17**) [12], 24-methyl-cholesta-5,24(28)-dien-3β,7β-diol (**18**) [13], 24-methyl-cholesta-5,24(28)-dien-3β,19-diol (**19**) [14], 24-methyl-cholesta-5,24(28)-dien-3β,7β,19-triol (**21**) [15], 7β-acetoxy-24-methyl-cholesta-5,24(28)-dien-3β,19-diol (**22**) [16], 7β-acetoxy-cholest-5-en-3β,19-diol (**23**) [15], 7-oxo-24-methyl-cholesta-5,24(28)-dien-3β,19-diol (**24**) [17], 24-methyl-cholesta-7,24(28)-dien-3β,5α,6β-triol (**25**) [18], and (22*E*,24*R*)-24-methyl-cholesta-7,22-dien-3β,5α,6β-triol (**26**) [18], previously isolated from various marine organisms, mainly soft corals of the genera *Litophyton* and *Nephthea*. Even though compound **4** has been isolated in the past, only a few characteristic ^{1}H NMR resonances have been reported. Analysis of its 1D and 2D spectra allowed for the full assignment of the ^{1}H and ^{13}C chemical shifts of compound **4** (Tables 1 and 2).

2.2. Evaluation of the Biological Activity of the Isolated Metabolites

Compounds **3–7**, **9–12**, **14–20**, and **22–26**, which were isolated in sufficient amounts, were evaluated in vitro in human tumor and non-cancerous cell lines for a number of biological activities, including cytotoxicity, anti-inflammatory, anti-angiogenic, and neuroprotective activity, as well as for their effect on androgen receptor (AR)-regulated transcription.

Initially, the cytotoxic activity of metabolites **3–7**, **9–12**, **14–20**, and **22–26** was evaluated in human cancer and normal cells after 72 h of treatment. Human cervical cancer (HeLa), human breast adenocarcinoma (MCF7) cell lines, and human normal fibroblasts (BJ) were used for the screening. Among the tested compounds, **22** and **23** strongly reduced the viability of cancer cells in the low micromolar range (Table 3), compounds **9–12**, **14–16**, **18**, and **24** showed moderate cytotoxic activity, while the remaining ten steroids were proven inactive. Most of the compounds with activity against cancer cells also showed cytotoxicity toward normal cells (BJ), except for compounds **11**, **12**, **14**, **15**, and **18**. Compared to cisplatin, these compounds exhibit a wide therapeutic window because of the absence of cytotoxicity on normal human fibroblasts. Moreover, metabolite **22** was proven more active against the HeLa cell line than the reference standard cisplatin.

Table 3. Cytotoxicity (IC_{50}; μM) of compounds **3–7**, **9–12**, **14–20**, and **22–26** against human cancer cell lines and fibroblasts after 72 h of treatment. Cisplatin was used as a positive control.

Compound	HeLa	MCF7	BJ
3	>50	>50	>50
4	>50	>50	>50
5	>50	>50	>50
6	>50	>50	>50
7	>50	>50	>50
9	25.9 ± 4.9	36.4 ± 5.7	18.3 ± 4.0
10	32.9 ± 6.8	32.7 ± 1.3	4.1 ± 1.9
11	18.8 ± 6.4	21.7 ± 1.4	>50
12	15.7 ± 2.0	25.3 ± 6.5	>50
14	>50	29.1 ± 5.0	>50
15	19.0 ± 4.3	18.9 ± 0.1	>50
16	32.9 ± 5.2	33.8 ± 1.0	23.2 ± 1.4
17	>50	>50	>50
18	22.6 ± 0.4	28.6 ± 5.3	>50
19	>50	>50	>50
20	>50	>50	>50
22	7.5 ± 0.1	8.9 ± 0.0	14.8 ± 5.8
23	12.0 ± 1.7	11.2 ± 0.5	14.5 ± 3.9
24	21.4 ± 2.0	31.7 ± 0.3	45.9 ± 2.8
25	>50	>50	>50
26	>50	>50	>50
cisplatin	11.4 ± 3.8	7.7 ± 1.7	6.9 ± 0.9

Subsequently, we examined whether the isolated steroids could influence angiogenesis or inflammation in vitro. Compounds **3–7**, **9–12**, **14–20**, and **22–26** and 2-methoxy-estradiol, which is known as an anti-angiogenic drug for the treatment of tumors and was used herein as positive control [19], were tested in the migration scratch and the tube formation assays using human umbilical vein endothelial cells (HUVEC). Only non-cytotoxic concentrations in HUVECs were used for the scratch assay. Metabolites **11**, **12**, **22**, and **23** partially inhibited HUVEC migration at 20 μM after 20 h of treatment (Figure 4), while in the tube formation assay, where HUVECs may form tube-like structures, no activity was observed for the tested compounds (data not shown). Thus, all tested steroids showed either little or no antiangiogenic activity.

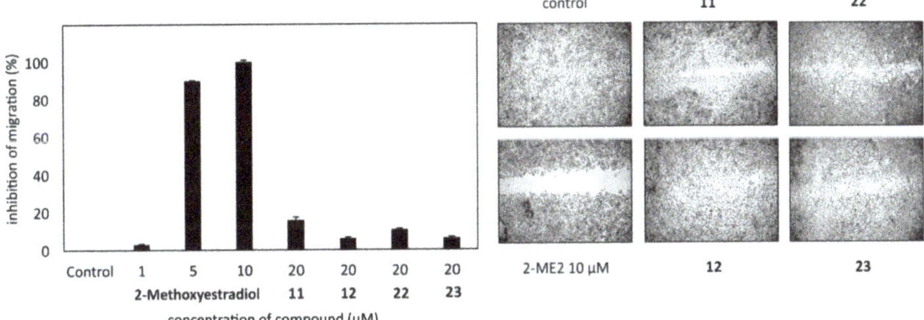

Figure 4. Compounds **11**, **12**, **22**, and **23** inhibited migration of HUVECs after 20 h of treatment at 20 µM. 2-Methoxy-estradiol was used as a positive control. The experiment was repeated three times in triplicates.

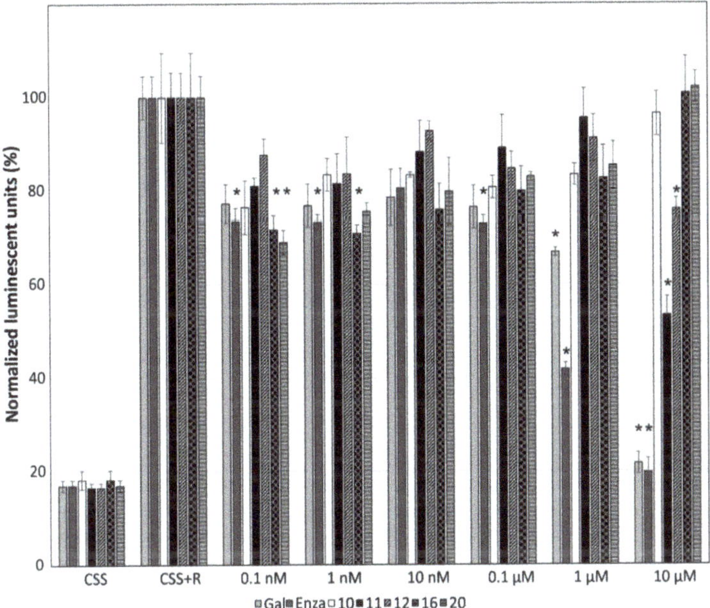

Figure 5. The influence of compounds **10**, **11**, **12**, **16**, and **20** on the androgen receptor-mediated transcription in the 22Rv1-ARE14 reporter cell line. Control cells were grown in charcoal-stripped serum medium (CSS). Cells were stimulated with either 1 nM methyltrienolone R1881 (R) or with the tested compounds in six different concentrations for 24 h in CSS. The luciferase activity was measured in the cell lysate. Enzalutamide (Enza) and galeterone (Gal) were used as positive controls. The experiment was repeated three times in triplicates. Bars with asterisk (*) are significantly different from the control (CSS+R) based on Tukey's multiple comparison test ($p \leq 0.05$).

The anti-inflammatory properties of metabolites **3–7**, **9–12**, **14–20**, and **22–26** were determined by measuring the levels of endothelial leukocyte adhesion molecule-1 (ELAM/E-selectin), which is a key molecular marker in the initiation of inflammation, expressed on the cell surface. Cell adhesion molecules (ICAM-1, VCAM-1, E-selectin) are significantly increased on the vascular endothelium activated by pro-inflammatory mediators (tumor necrosis factor α, TNFα) as a crucial step for the extravasation of leukocytes into inflamed tissue [20]. TNFα stimulates NFκB (nuclear factor kappa-light-chain-enhancer of activated B cells) and thus E-selectin (CD62E, ELAM). Endothelial cells were pre-treated for 30 min with the tested compounds and then activated with TNFα for 4 h. Curcumin, which inhibits activation of NF-κB and thus inhibits expression and activity of the COX-2 gene induced by TNFα [21], was used as a positive control, decreasing ELAM production to 25% at 10 μM. None of the tested compounds decreased the levels of ELAM (Supplementary Materials Figure S1).

We have previously observed that several steroids with bulky or long side chains, such as galeterone derivatives or cholestanes, can also inhibit AR [22,23]. Therefore, we further analyzed the influence of the isolated steroids on AR-mediated transcription. Compounds **3–7**, **9–12**, **14–20**, and **22–26** in six different concentrations were evaluated on the reporter cell line 22Rv1-ARE14 for 24 h. Galeterone and enzalutamide, which were used as positive controls, showed a strong dose-dependent reduction of AR-regulated transcription (22% and 20% inhibition at 10 μM, respectively). Compounds **11**, **12**, **16**, and **20** showed clearly reduced luciferase activity in a dose-dependent manner, already at submicromolar concentrations (Figure 5). The strongest inhibition of AR was displayed by compound **11** at 10 μM. Surprisingly, steroids **10**, **16**, and **20** showed increased inhibition of AR (or decreased the luciferase activity) with decreasing concentrations (Figure 5). All other steroids tested inhibited AR by no more than 20% (data not shown).

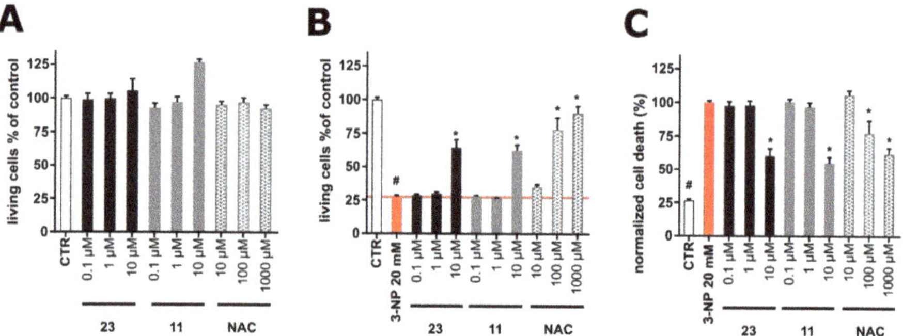

Figure 6. (**A**) Cytotoxicity of compounds **11** and **23** in human neuron-like SH-SY5Y cells after 48 h of treatment. All results are presented as mean ± standard error of the mean (SEM) from at least three independent experiments in triplicates. (**B**) Neuroprotective activity of compounds **11** and **23** in the 3-nitropropionic acid (3-NPA)-induced model of Huntington's disease on human neuron-like SH-SY5Y cells after 48 h of treatment. (**C**) Cell death of neuron-like SH-SY5Y induced by 3-NPA and the protective effect of compounds **11** and **23** after 48 h. The results are presented as mean ± standard error of the mean (SEM) from triplicates in four independent experiments (*n* = 4). *N*-acetylcysteine (NAC) was used as a positive control. * *p* compared with vehicle with 20 mM 3-NPA, # *p* compared with vehicle without 20 mM 3-NPA.

Since anti-androgens have shown neuroprotective effects in the in vivo model of Huntington's disease [24], selected compounds that showed cytotoxic activity and moderate activity in the migration scratch assay were tested for their cytotoxicity on differentiated human SH-SY5Y cells (neuron-like cells). Specifically, based on the combined results obtained for compounds **3–7**, **9–12**, **14–20**, and **22–26** in the cytotoxicity assay, the anti-inflammatory activity assay, the migration scratch assay, and the tube formation assay in HUVECs, as well as in the reporter assay with AR, compounds **11** and **23** were selected as the two most active compounds that were subsequently evaluated for their neuroprotective activity. As shown in Figure 6A, compounds **11** and **23**, as well as *N*-acetylcystein (NAC) that was used as a positive control, did not show cytotoxic, but rather stimulatory activity. In order to evaluate the potential neuroprotective effect of the compounds, neuroblastoma cell line SH-SY5Y was differentiated for 48 h and further exposed to 20 mM 3-nitropropionic acid (3-NPA) as an agent mimicking Huntington's disease in vitro [25]. 3-NPA was used alone or in co-treatment with the tested compounds at concentrations of 0.1–10 µM. As shown in Figure 6B, 3-NPA caused dramatic (approx. 70%) decrease in cell viability determined by the Calcein AM assay. The positive control (NAC) showed partial (cell viability 77.9 ± 8.99% at 100 µM) or almost complete (cell viability 90.2 ± 5.28% at 1000 µM) protection of cells from the negative effect of 3-NPA. Compounds **11** and **23** showed significant protective effects at 10 µM (cell viability 64.4 ± 6.02% for **11** and 62.5 ± 3.95% for **23**), comparable to 100 µM of the positive control NAC. Being encouraged by the promising results, we further analyzed the protective effects of **11** and **23** using an orthogonal method (propidium iodide (PI) assay) to verify their activity. In general, PI as a positively charged dye is associated with an increase of cell damage or death, since it penetrates cells with cell-disrupted membranes [26]. Within the 3-NPA model, its toxic effect was considered as 100% of the PI signal and thus reduction of cell death was determined. As shown in Figure 6C, compounds **11** and **23** significantly reduced cell death (maximal effect at 10 µM) in a manner similar to NAC (100 µM, 77.2 ± 9.53%; 1000 µM, 62.1 ± 4.19%), thus confirming the protective effects of **11** and **23** observed in the viability assay.

3. Materials and Methods

3.1. General Experimental Procedures

Optical rotations were measured on a Krüss polarimeter (A. KRÜSS Optronic GmbH, Hamburg, Germany) equipped with a 0.5 dm cell. UV spectra were recorded on a Lambda 40 UV/Vis spectrophotometer (Perkin Elmer Ltd., Beaconsfield, UK). IR spectra were obtained on an Alpha II FTIR spectrometer (Bruker Optik GmbH, Ettlingen, Germany). Low-resolution EI mass spectra were measured on a Thermo Electron Corporation DSQ mass spectrometer (Thermo Fisher Scientific, Bremen, Germany). High-resolution APCI or ESI mass spectra were measured on a LTQ Orbitrap Velos mass spectrometer (Thermo Fisher Scientific, Bremen, Germany). NMR spectra were recorded on Bruker AC 200, DRX 400, and Avance NEO 950 (Bruker BioSpin GmbH, Rheinstetten, Germany) and Varian 600 (Varian, Inc., Palo Alto, CA, USA) spectrometers. Chemical shifts are given on the δ (ppm) scale with reference to the solvent signals. The 2D NMR experiments (HSQC, HMBC, COSY, NOESY) were performed using standard Bruker or Varian pulse sequences. Column chromatography separations were performed with Kieselgel 60 (Merck, Darmstadt, Germany). HPLC separations were conducted on a Waters 600 liquid chromatography pump equipped with a Waters 410 differential refractometer (Waters Corporation, Milford, MA, USA), using a Kromasil 100 C_{18} (25 cm × 8 mm i.d.) column (MZ-Analysentechnik GmbH, Mainz, Germany). TLC were performed with Kieselgel 60 F_{254} aluminum plates (Merck, Darmstadt, Germany) and spots were detected after spraying with 25% H_2SO_4 in MeOH reagent and heating at 100 °C for 1 min.

3.2. Biological Material

Specimens of *S. polydactyla* were hand-picked by SCUBA diving at a depth of 10 m from the reefs near the National Institute of Oceanography and Fisheries (NOIF), Hurghada, Egypt (GPS coordinates 27°17′06″N, 33°46′24″E) in June 2015 and transported to the laboratory in ice chests, where they were stored at −20 °C until analyzed. A voucher specimen has been deposited at the animal collection of NOIF in Hurghada and the animal collection of the Section of Pharmacognosy and Chemistry of Natural Products, Department of Pharmacy, National and Kapodistrian University of Athens (ATPH/MP0533).

3.3. Extraction and Isolation

Specimens of the freeze-dried gorgonian (119.9 g) were exhaustively extracted with mixtures of CH_2Cl_2/MeOH (2:1) at room temperature. Evaporation of the solvents under vacuum afforded a dark green residue (18.5 g) which was submitted to vacuum column chromatography on silica gel using cHex with increasing amounts of EtOAc followed by EtOAc with increasing amounts of MeOH as eluent to yield 9 fractions (A–I). Fraction B (2.5 g, 30–50% EtOAc in cHex) was fractionated by vacuum column chromatography on silica gel using mixtures of cHex/EtOAc of increasing polarity as mobile phase to yield 6 fractions (B1–B6). Fraction B6 (0.5 g, 20–30% EtOAc in cHex) was further fractionated by gravity column chromatography on silica gel using mixtures of cHex/EtOAc of increasing polarity as mobile phase to afford 10 fractions (B6a–B6j). Fractions B6e, B6g, and B6h were subjected repeatedly to reversed-phase HPLC, using MeOH/H$_2$O (100:0 and 98:2) as eluent to yield compounds **3** (9.9 mg), **4** (8.8 mg), **5** (1.8 mg), **6** (5.9 mg), **8** (0.8 mg), and **17** (14.0 mg). Fraction C (2.5 g, 60–70% EtOAc in cHex) was submitted to vacuum column chromatography on silica gel using mixtures of cHex/EtOAc of increasing polarity as mobile phase to afford 9 fractions (C1–C9). Fractions C5, C6, C7, and C8 were subjected repeatedly to reversed-phase HPLC using MeOH/H$_2$O (100:0, 98:2, and 97:3) as eluent to afford compounds **1** (0.7 mg), **2** (0.3 mg), **7** (6.3 mg), **9** (5.7 mg), and **10** (5.7 mg). Fraction D (0.9 g, 80–90% EtOAc in cHex) was fractionated by gravity column chromatography on silica gel using cHex with increasing amounts of Me$_2$CO as mobile phase to afford 18 fractions (D1–D18). Fractions D8, D9, D11, and D12 were subjected repeatedly to reversed-phase HPLC using MeOH/H$_2$O (100:0 and 98:2) as eluent to yield compounds **11** (4.0 mg), **12** (2.2 mg), **13** (1.3 mg), **14** (1.1 mg), **15** (9.4 mg), and **16** (2.3 mg). Fraction E (0.23 g, 100% EtOAc) was fractionated by gravity column chromatography on silica gel using cHex with increasing amounts of Me$_2$CO as mobile phase to afford 10 fractions (E1–E10). Fractions E3, E5, and E6 were subjected repeatedly to reversed-phase HPLC using mixtures of MeOH/H$_2$O (100:0 and 97:3) as eluent to yield compounds **18** (1.7 mg), **19** (17.5 mg), **20** (2.0 mg), **22** (2.1 mg), and **23** (1.0 mg). Fraction G (1.7 g, 50% MeOH in EtOAc) was separated by vacuum column chromatography on silica gel using cHex with increasing amounts of EtOAc and EtOAc with increasing amounts of MeOH to yield 7 fractions (G1–G7). Fractions G5 and G7 were subjected repeatedly to reversed-phase HPLC using MeOH/H$_2$O (100:0 and 97:3) as eluent to yield compounds **24** (6.9 mg), **25** (2.0 mg), and **26** (3.1 mg). Fraction I (5.1 g, 100% MeOH) was subjected to vacuum column chromatography on silica gel using cHex with increasing amounts of EtOAc followed by EtOAc with increasing amounts of MeOH to afford 11 fractions (I1–I11). Fractions I6 and I7 were combined and subjected repeatedly to reversed-phase HPLC using MeOH (100%) as eluent to yield compound **21** (2.3 mg).

(9*E*,22*E*,24*R*)-3β-Hydroxy-4α,24-dimethyl-8,9-seco-5α-cholesta-9(11),22-dien-8-one (**1**): White amorphous solid; ^{1}H and ^{13}C NMR data, see Tables 1 and 2; HR-APCIMS *m/z* 429.3724 [M + H]$^+$ (calcd. for C$_{29}$H$_{49}$O$_2$, 429.3727).

(9*E*)-3β-Hydroxy-4α,24-dimethyl-8,9-seco-5α-cholesta-9(11),24(28)-dien-8-one (**2**): White amorphous solid; ^{1}H NMR data, see Table 1; HR-APCIMS *m/z* 429.3725 [M + H]$^+$ (calcd. for C$_{29}$H$_{49}$O$_2$, 429.3727).

(22*E*,24*R*)-4α,24-Dimethyl-5α-cholest-22-en-3β,8β-diol (**6**): White amorphous solid; $[\alpha]_D^{20}$ + 68 (*c* 0.25, CHCl$_3$); UV (CHCl$_3$) λ_{max} (log ε) 206 (2.85); IR (thin film) ν_{max} 3440, 2954, 2856, 1466, 1384, 964 cm^{-1}; ^{1}H and ^{13}C NMR data, see Tables 1 and 2; HR-APCIMS *m/z* 413.3773 [M − H$_2$O + H]$^+$ (calcd. for C$_{29}$H$_{49}$O, 413.3778).

(23*E*)-22α,28-Epidioxy-4α,24-dimethyl-5α-cholest-23-en-3β,8β-diol (**8**): White amorphous solid; $[\alpha]_D^{20}$ −50 (*c* 0.02, CHCl$_3$); UV (CHCl$_3$) λ_{max} (log ε) 206 (3.94), 225 (3.84); IR (thin film) ν_{max} 3458, 2927, 2861, 1723, 1462, 1378, 1243, 1013 cm^{-1}; ^{1}H and ^{13}C NMR data, see Tables 1 and 2; HR-APCIMS *m/z* 459.3475 [M − H]$^-$ (calcd. for C$_{29}$H$_{47}$O$_4$, 459.3480).

(23*E*)-22α,28-Epidioxy-4α,24-dimethyl-5α-cholest-23-en-3β,8β,11β-triol (**13**): White amorphous solid; $[\alpha]_D^{20}$ +30 (*c* 0.1, CHCl$_3$); UV (CHCl$_3$) λ_{max} (log ε) 206 (3.26), 230 (2.97); IR (thin film) ν_{max} 3403, 2923, 2873, 1725, 1450, 1377, 1261, 1045 cm^{-1}; ^{1}H and ^{13}C NMR data, see Tables 1 and 2; HR-APCIMS *m/z* 475.3424 [M − H]$^-$ (calcd. for C$_{29}$H$_{47}$O$_5$, 475.3429).

(22*E*)-24-Methyl-cholesta-5,22,24(28)-trien-3β,19-diol (**20**): White amorphous solid; $[\alpha]_D^{20}$ −133 (*c* 0.03, CHCl$_3$); UV (CHCl$_3$) λ_{max} (log ε) 206 (3.61), 232 (3.76); IR (thin film) ν_{max} 3433, 2919, 2857, 1462, 1377, 1261, 1037 cm^{-1}; ^{1}H and ^{13}C NMR data, see Tables 1 and 2; HR-ESIMS *m/z* 411.3271 [M − H]$^-$ (calcd. for C$_{28}$H$_{43}$O$_2$, 411.3269).

3.4. Cell Culture

The tested compounds were dissolved in DMSO to afford 10 mM stock solutions. Human cervical carcinoma (HeLa) and human breast adenocarcinoma (MCF7) cell lines were purchased from European Collection of Authenticated Cell Cultures (ECACC, Salisbury, UK) and cultivated in Dulbecco's Modified Eagle Medium (DMEM) (Merck, Darmstadt, Germany), as previously reported [27]. Human umbilical vein endothelial cells (HUVECs) were a kind gift of Prof. Jitka Ulrichová (Faculty of Medicine and Dentistry, Palacky University, Olomouc, Czech Republic). The cultivation and assay was performed in endothelial cell proliferation medium (ECPM, Provitro, Berlin, Germany) [23]. The 22Rv1-ARE14 reporter cell line [28] was a generous gift of Prof. Zdeněk Dvořák (Department of Cell Biology and Genetics, Palacky University). The 22Rv1-ARE14 cell line was grown in Roswell Park Memorial Institute 1640 Medium (RPMI 1640) [22]. All cells were maintained in a humidified CO_2 incubator at 37 °C using the standard trypsinization procedure twice or three times a week. The SH-SY5Y human neuroblastoma cell line (ECACC, Salisbury, UK) was cultivated in DMEM and Ham's F12 Nutrient Mixture (DMEM:F12, 1:1), as previously described [25]. Cells were used up to twenty passages. All trans-retinoic acid (10 μM) in 1% FBS DMEM/F12 medium was added to SH-SY5Y cells to achieve differentiation conditions [29,30], to reach longer neurites and reduced proliferation (48 h). All cells were maintained in a humidified CO_2 incubator at 37 °C using the standard trypsinization procedure twice or three times a week.

3.5. Evaluation of Cytotoxicity

In cytotoxicity assays, cancer cells were treated with six different concentrations of each tested compound for 72 h. Cells were stained with resazurin and IC_{50} values were calculated as previously reported [26]. Triplicates from at least three independent experiments were used. For the ELAM assay, the Calcein AM (Molecular Probes, Invitrogen, Karlsruhe, Germany) cytotoxicity assay, which assessed HUVEC viability after 4 h treatment, was used as previously described [31].

3.6. Cell-Surface ELISA CD62E (E-Selectin, ELAM)

An enzyme-linked immunosorbent assay (ELISA) was used to detect the levels of the cell adhesion molecule ELAM in HUVEC cells after 30 min of incubation with the tested compounds and 4 h of stimulation with TNFα, as previously described [31].

3.7. Migration Scratch Assay

The scratch test was performed with HUVEC cells and evaluated after 20 h of treatment, as previously reported [23].

3.8. AR-Transcriptional Reporter Assay

AR-transcriptional reporter assays were performed on 22Rv1-ARE14 cells after 24 h of incubation, as previously described [22].

3.9. SH-SY5Y Cell Treatments and Evaluation of Cell Viability/Cytotoxicity

Cell viability of neuron-like SH-SY5Y cells growing in 96-well plates (7000 cells/well) 24 h after treatment was evaluated using the Calcein AM assay [27], with minor modification of Calcein AM concentration (0.75 μM). Cell death of SH-SY5Y cells (20,000 cells/well) was determined using the propidium iodide (PI) assay according to Stone et al. with a slight modification [32]. Briefly, PI solution in PBS is added to cell medium to reach concentration 1 μg/mL, incubated for 15 min at room temperature and quantified at 535/617 nm (excitation/emission) by Infinite M200 Pro reader (Tecan, Austria). The resulting fluorescence of 3-NPA toxin was considered as 100% cell death. In the assays, neuron-like cells were treated with the tested compounds in 0.1–10 μM concentration range for 48 h. DMSO-treated cells (≤0.1% *v/v*) were used as healthy controls.

3.10. Statistical Analysis

All data are expressed as mean ± SD or SEM. Data were evaluated by non-parametric Kruskal–Wallis test followed by post-hoc Mann–Whitney test (Figure 6) with sequential Bonferroni correction of *p*-values or ANOVA followed by Tukey's multiple comparison test (Figure 5) using the PAST (version 1.97) software package [33]. *p*-values < 0.05 were considered statistically significant.

4. Conclusions

The chemical analysis of the organic extract of the soft coral *S. polydactyla* collected from the Hurghada reef in the Red Sea resulted in the isolation of 26 steroids, with six of them (**1**, **2**, **6**, **8**, **13**, and **20**) being new natural products. Among them, **1** and **2** display the rare 8,9-seco-cholestane steroidal nucleus. To the best of our knowledge, dictyoneolone, isolated from a sponge of the genus *Dictyonella*, is the only previously reported 4-methyl-8,9-seco-cholestane derivative [34]. Evaluation of cytotoxic, anti-inflammatory, anti-angiogenic, and neuroprotective activity of the majority of the isolated metabolites that were isolated in adequate quantities revealed significant cytotoxicity in the low micromolar range against the HeLa and MCF7 cancer cell lines for compounds **22** and **23**, while compounds **11**, **12**, **22**, and **23** inhibited the migration of endothelial cells at 20 µM. Most of the compounds with activity against cancer cells also showed cytotoxicity toward normal cells (BJ), except for compounds **11**, **12**, **14**, **15**, and **18**. Compared to cisplatin, these compounds have therefore a broad therapeutic window due to low or zero cytotoxicity on normal human fibroblasts. Moreover, metabolite **22** was more active against HeLa cells than cisplatin used as positive control. Furthermore, the effect of the isolated metabolites on AR-regulated transcription was evaluated in vitro in human tumor and non-cancerous cells, with compound **11** exhibiting the strongest inhibition of AR at 10 µM. It is worth-noting that metabolites **10**, **16**, and **20** displayed increased inhibition of AR with decreasing concentrations. In addition, compounds **11** and **23** showed neuroprotective activity on neuron-like SH-SY5Y cells.

Supplementary Materials: The following are available online at http://www.mdpi.com/1660-3397/18/12/632/s1. Figure S1: Visual representation of the evaluation of the activity of the tested metabolites against ELAM. Figures S2–S29: 1D and 2D NMR and HR-MS spectra of compounds **1**, **2**, **6**, **8**, **13**, and **20**.

Author Contributions: Conceptualization, M.S., E.I., and V.R.; methodology, L.R., M.K., G.G., M.S., E.I., and V.R.; formal analysis, M.A.T., L.R., and E.I.; investigation, M.A.T., L.R., M.K., G.G., A.M., M.S., E.I., and V.R.; resources, M.S., E.I., and V.R.; writing—original draft preparation, M.A.T., L.R., M.K., G.G., and E.I.; writing—review and editing, L.R., A.M.E., M.S., E.I., and V.R.; visualization, M.A.T., L.R., and E.I.; supervision, M.S., E.I., and V.R.; project administration, M.S., E.I., and V.R.; funding acquisition, M.S., E.I., and V.R. All authors have read and agreed to the published version of the manuscript.

Funding: This work was supported by the research projects MARINOVA (grant number 70/3/14684, V.R.) and BioNP (grant number 70/3/14685, E.I.). Biological activity evaluation was supported by the grants of Czech Science Foundation 19-01383S, 20-15621S and IGA_PrF_2020_021.

Acknowledgments: M.A.T. acknowledges support by the mission sector of the Ministry of Higher Education of the Arab Republic of Egypt (Egyptian Cultural Bureau in Athens), the Directorate of Education and Cultural Affairs, Ministry of Foreign Affairs of Greece and the non-profit organization "Kleon Tsetis". We thank Jitka Ulrichová for the kind gift of HUVEC cells and Zdeněk Dvořák for the kind gift of 22Rv1-ARE14 cells. Authors also thank Anežka Šindlerová and Veronika Górová for excellent technical assistance. Assistance on the acquisition of some NMR data by Lukasz Jaremko (King Abdullah University of Science and Technology, Saudi Arabia) is gratefully acknowledged.

References

1. Kotb, M.M.A.; Hanafy, M.H.; Rirache, H.; Matsumura, S.; Al-Sofyani, A.A.; Ahmed, A.G.; Bawazir, G.; Al-Horani, F.A. Status of coral reefs in the Red Sea and Gulf of Aden region. In *Status of Coral Reefs of the World: 2008*; Wilkinson, C., Ed.; Global Coral Reef Monitoring Network and Reef and Rainforest Research Centre: Townsville, Australia, 2008; pp. 67–78.
2. Carroll, A.R.; Copp, B.R.; Davis, R.A.; Keyzers, R.A.; Prinsep, M.R. Marine natural products. *Nat. Prod. Rep.* **2020**, *37*, 175–223. [CrossRef] [PubMed]
3. MarinLit. A Database of the Marine Natural Products Literature. Available online: http://pubs.rsc.org/marinlit/ (accessed on 30 September 2020).
4. Končić, M.; Ioannou, E.; Sawadogo, W.; Abdel-Razik, A.; Vagias, C.; Diederich, M.; Roussis, V. 4*a*-Methylated steroids with cytotoxic activity from the soft coral *Litophyton mollis*. *Steroids* **2016**, *115*, 130–135. [CrossRef] [PubMed]
5. Yu, S.; Deng, Z.; van Ofwegen, L.; Proksch, P.; Lin, W. 5,8-Epidioxysterols and related derivatives from a Chinese soft coral *Sinularia flexibilis*. *Steroids* **2006**, *71*, 955–959. [CrossRef] [PubMed]
6. Kokke, W.; Bohlin, L.; Fenical, W.; Djerassi, C. Novel dinoflagellate 4*α*-methylated sterols from four caribbean gorgonians. *Phytochemistry* **1982**, *21*, 881–887. [CrossRef]
7. Kobayashi, M.; Ishizaka, T.; Mitsuhashi, H. Marine sterols X. Minor constituents of the sterols of the soft coral *Sarcophyton glaucum*. *Steroids* **1982**, *40*, 209–221. [CrossRef]
8. Mehta, G.; Venkateswarlu, Y.; Rama, R.M.; Uma, R. A novel 4*α*-methyl sterol from the soft coral *Nephthea chabroli*. *J. Chem. Res.* **1999**, *23*, 628–629. [CrossRef]
9. Bortolotto, M.; Braekman, J.; Daloze, D.; Tursch, B. Chemical studies of marine invertebrates. XXIX. 4*α*-methyl-3*β*,8*β*-dihydroxy-5*α*-ergost-24(28)-en-23-one, a novel polyhydroxygenated sterol from the soft coral *litophyton viridis*. *Steroids* **1977**, *30*, 159–164. [CrossRef]
10. Cheng, S.; Huang, Y.; Wen, Z.; Hsu, C.; Wang, S.; Dai, C.; Duh, C. New 19-oxygenated and 4-methylated steroids from the formosan soft coral *Nephthea chabroli*. *Steroids* **2009**, *74*, 543–547. [CrossRef]
11. Huang, Y.; Wen, Z.; Wang, S.; Hsu, C.; Duh, C. New anti-inflammatory 4-methylated steroids from the formosan soft coral *Nephthea chabroli*. *Steroids* **2008**, *73*, 1181–1186. [CrossRef]
12. Viegelmann, C.; Parker, J.; Ooi, T.; Clements, C.; Abbott, G.; Young, L.; Kennedy, J.; Dobson, A.; Edrada-Ebel, R. Isolation and identification of antitrypanosomal and antimycobacterial active steroids from the sponge *Haliclona simulans*. *Mar. Drugs* **2014**, *12*, 2937–2952. [CrossRef]
13. Riccardis, F.; Minale, L. Marine sterols side-chain-oxygenated sterols, possibly of abiotic origin, from the new caledonian sponge *stelodoryx chlorophylla*. *J. Nat. Prod.* **1993**, *56*, 282–287. [CrossRef]
14. Iguchi, K.; Saitou, S.; Yamada, Y. Novel 19-oxygenated sterols from the okinawan soft coral *Litophyton viridis*. *Chem. Pharm. Bull.* **1989**, *37*, 2553–2554. [CrossRef]
15. Cheng, S.; Dai, C.; Duh, C. New 4-methylated and 19-oxygenated steroids from the formosan soft coral *Nephthea erecta*. *Steroids* **2007**, *72*, 653–659. [CrossRef] [PubMed]
16. Ellithey, M.; Lall, N.; Hussein, A.; Meyer, D. Cytotoxic, cytostatic and HIV-1 PR inhibitory activities of the soft coral *Litophyton arboretum*. *Mar. Drugs* **2013**, *11*, 4917–4936. [CrossRef]
17. Duh, C.; Wang, S.; Chu, M.; Sheu, J. Cytotoxic sterols from the soft coral *Nephthea erecta*. *J. Nat. Prod.* **1998**, *61*, 1022–1024. [CrossRef]
18. Iorizzi, M.; Minale, L.; Riccio, R. Polar steroids from the marine scallop *Patinopecten yessoensis*. *J. Nat. Prod.* **1988**, *51*, 1098–1103. [CrossRef]
19. Sattler, M.; Quinnan, L.R.; Pride, Y.B.; Gramlich, J.L.; Chu, S.C.; Even, G.C.; Kraeft, S.-K.; Chen, L.B.; Salgia, R. 2-Methoxyestradiol alters cell motility, migration, and adhesion. *Blood* **2003**, *102*, 289–296. [CrossRef]
20. Trepels, T.; Zeiher, A.M.; Fichtlscherer, S. The endothelium and inflammation. *Endothelium* **2006**, *13*, 423–429. [CrossRef]
21. Zhang, F.; Altorki, N.K.; Mestre, J.R.; Subbaramaiah, K.; Dannenberg, A.J. Curcumin inhibits cyclooxygenase-2 transcription in bile acid- and phorbol ester-treated human gastrointestinal epithelial cells. *Carcinogenesis* **1999**, *20*, 445–451. [CrossRef]
22. Jorda, R.; Řezníčková, E.; Kiełczewska, U.; Maj, J.; Morzycki, J.W.; Siergiejczyk, L.; Bazgier, V.; Berka, K.; Rárová, L.; Wojtkielewicz, A. Synthesis of novel galeterone derivatives and evaluation of their in vitro activity against prostate cancer cell lines. *Eur. J. Med. Chem.* **2019**, *179*, 483–492. [CrossRef]

23. Rárová, L.; Sedlák, D.; Oklestkova, J.; Steigerová, J.; Liebl, J.; Zahler, S.; Bartůněk, P.; Kolář, Z.; Kohout, L.; Kvasnica, M.; et al. The novel brassinosteroid analog BR4848 inhibits angiogenesis in human endothelial cells and induces apoptosis in human cancer cells in vitro. *J. Steroid Biochem. Mol. Biol.* **2018**, *178*, 263–271.

24. Calderon Guzman, D.; Bratoeff, E.; Chávez-Riveros, A.; Osnaya, N.; Barragan, G.; Hernandez Garcia, E.; Olguín, H.; Garcia, E. Effect of two antiandrogens as protectors of prostate and brain in a Huntington's animal model. *Anticancer Agents Med. Chem.* **2014**, *14*, 1293–1301. [CrossRef] [PubMed]

25. Colle, D.; Santos, D.; Hartwig, J.; Godoi, M.; Engel, D.; de Bem, A.; Braga, A.; Farina, M. Succinobucol, a lipid-lowering drug, protects against 3-nitropropionic acid-induced mitochondrial dysfunction and oxidative stress in SH-SY5Y cells via upregulation of glutathione levels and glutamate cysteine ligase activity. *Mol. Neurobiol.* **2016**, *53*, 1280–1295. [CrossRef] [PubMed]

26. Dengler, W.A.; Schulte, J.; Berger, D.P.; Mertelsmann, R.; Fiebig, H.H. Development of a propidium iodide fluorescence assay for proliferation and cytotoxicity assays. *Anticancer Drugs* **1995**, *6*, 522–532. [CrossRef]

27. Rárová, L.; Steigerová, J.; Kvasnica, M.; Bartůněk, P.; Křížová, K.; Chodounská, H.; Kolář, Z.; Sedlák, D.; Oklestkova, J.; Strnad, M. Structure activity relationship studies on cytotoxicity and the effects on steroid receptor of AB-functionalized cholestanes. *J. Steroid Biochem. Mol. Biol.* **2016**, *159*, 154–169. [CrossRef]

28. Bartonkova, I.; Novotna, A.; Dvorak, Z. Novel stably transfected human reporter cell line AIZ-AR as a tool for an assessment of human androgen receptor transcriptional activity. *PLoS ONE* **2015**, *10*, e0121316. [CrossRef]

29. Cheung, Y.-T.; Lau, W.K.-W.; Yu, M.-S.; Lai, C.S.-W.; Yeung, S.-C.; So, K.-F.; Chang, R.C.-C. Effects of all-trans-retinoic acid on human SH-SY5Y neuroblastoma as in vitro model in neurotoxicity research. *Neurotoxicology* **2009**, *30*, 127–135. [CrossRef]

30. Dwane, S.; Durack, E.; Kiely, P.A. Optimising parameters for the differentiation of SH-SY5Y cells to study cell adhesion and cell migration. *BMC Res. Notes* **2013**, *6*, 366. [CrossRef]

31. Morrogh-Bernard, H.C.; Foitová, I.; Yeen, Z.; Wilkin, P.; de Martin, R.; Rárová, L.; Doležal, K.; Nurcahyo, W.; Olšanský, M. Self-medication by orang-utans (*Pongo pygmaeus*) using bioactive properties of *Dracaena cantleyi*. *Sci. Rep.* **2017**, *7*, 16653. [CrossRef]

32. Stone, W.L.; Qui, M.; Smith, M. Lipopolysaccharide enhances the cytotoxicity of 2-chloroethyl ethyl sulfide. *BMC Cell Biol.* **2003**, *4*, 1. [CrossRef]

33. Hammer, O.; Harper, D.A.T.; Ryan, P.D. PAST: Paleontological statistics software package for education and data analysis. *Palaeontol. Electron.* **2001**, *4*, 9.

34. Woo, J.-K.; Yun, J.-H.; Ahn, S.; Sim, C.J.; Noh, M.; Oh, D.-C.; Oh, K.-B.; Shin, J. Dictyoneolone, a B/C ring juncture-defused steroid from a *Dictyonella* sp. sponge. *Tetrahedron Lett.* **2018**, *59*, 2021–2024. [CrossRef]

Publisher's Note: MDPI stays neutral with regard to jurisdictional claims in published maps and institutional affiliations.

 *marine drugs*

Article

New Cytotoxic Natural Products from the Red Sea Sponge *Stylissa carteri*

Reda F. A. Abdelhameed [1,†], Eman S. Habib [1,†], Nermeen A. Eltahawy [1], Hashim A. Hassanean [1], Amany K. Ibrahim [1], Anber F. Mohammed [2], Shaimaa Fayez [3,4], Alaa M. Hayallah [2,5], Koji Yamada [6], Fathy A. Behery [7,8], Mohammad M. Al-Sanea [9], Sami I. Alzarea [10], Gerhard Bringmann [3], Safwat A. Ahmed [1,*] and Usama Ramadan Abdelmohsen [11,12,*]

[1] Department of Pharmacognosy, Faculty of Pharmacy, Suez Canal University, Ismailia 41522, Egypt; omarreda_70@yahoo.com (R.F.A.A.); emy_197@hotmail.com (E.S.H.); nermeenazmy25@gmail.com (N.A.E.); hasanean2000@yahoo.com (H.A.H.); am_kamal66@yahoo.com (A.K.I.)

[2] Department of Pharmaceutical Organic Chemistry, Faculty of Pharmacy, Assiut University, Assiut 71526, Egypt; anber_pharm_2006@yahoo.com (A.F.M.); alaa_hayalah@yahoo.com (A.M.H.)

[3] Institute of Organic Chemistry, University of Würzburg, Am Hubland, 97074 Würzburg, Germany; shaimaa.seaf@uni-wuerzburg.de (S.F.); bringmann@chemie.uni-wuerzburg.de (G.B.)

[4] Department of Pharmacognosy, Faculty of Pharmacy, Ain-Shams University, Cairo 11566, Egypt

[5] Department of Pharmaceutical Chemistry, Faculty of Pharmacy, Deraya University, New Minia 61111, Egypt

[6] Graduate School of Biomedical Sciences, Nagasaki University, Bunkyo-machi 1-14, Nagasaki 852-8521, Japan; kyamada@nagasaki-u.ac.jp

[7] Department of Pharmacognosy, Faculty of Pharmacy, Mansoura University, Mansoura 35516, Egypt; fathybehery@yahoo.com

[8] Department of Pharmaceutical Sciences, College of Pharmacy, Riyadh Elm University, Riyadh 11681, Saudi Arabia

[9] Department of Pharmaceutical Chemistry, College of Pharmacy, Jouf University, Aljouf 72341, Saudi Arabia; mohmah80@gmail.com

[10] Department of Pharmacology, College of Pharmacy, Jouf University, Aljouf 72341, Saudi Arabia; samisz@ju.edu.sa

[11] Department of Pharmacognosy, Faculty of Pharmacy, Deraya University, New Minia 61111, Egypt

[12] Department of Pharmacognosy, Faculty of Pharmacy, Minia University, Minia 61519, Egypt

* Correspondence: safwat_aa@yahoo.com (S.A.A.); usama.ramadan@mu.edu.eg (U.R.A.)

† Equal contributions: Eman S. Habib and Reda F. A. Abdelhameed as first author.

Received: 10 April 2020; Accepted: 27 April 2020; Published: 3 May 2020

Abstract: Bioactivity-guided isolation supported by LC-HRESIMS metabolic profiling led to the isolation of two new compounds, a ceramide, stylissamide A (**1**), and a cerebroside, stylissoside A (**2**), from the methanol extract of the Red Sea sponge *Stylissa carteri*. Structure elucidation was achieved using spectroscopic techniques, including 1D and 2D NMR and HRMS. The bioactive extract's metabolomic profiling showed the existence of various secondary metabolites, mainly oleanane-type saponins, phenolic diterpenes, and lupane triterpenes. The in vitro cytotoxic activity of the isolated compounds was tested against two human cancer cell lines, MCF-7 and HepG2. Both compounds, **1** and **2**, displayed strong cytotoxicity against the MCF-7 cell line, with IC_{50} values at 21.1 ± 0.17 µM and 27.5 ± 0.18 µM, respectively. They likewise showed a promising activity against HepG2 with IC_{50} at 36.8 ± 0.16 µM for **1** and IC_{50} 30.5 ± 0.23 µM for **2** compared to the standard drug cisplatin. Molecular docking experiments showed that **1** and **2** displayed high affinity to the SET protein and to inhibitor 2 of protein phosphatase 2A (I2PP2A), which could be a possible mechanism for their cytotoxic activity. This paper spreads light on the role of these metabolites in holding fouling organisms away from the outer surface of the sponge, and the potential use of these defensive molecules in the production of novel anticancer agents.

Keywords: LC-HRESIMS; *Stylissa carteri*; ceramide; cerebroside; docking; cytotoxic activity

1. Introduction

Marine environments have proven to be an important source of unique chemical entities with a wide range of biological activities [1–6]. Owing to its biodiversity and seasonal variations in air and water temperatures, the Red Sea is one of the most important areas for marine research. Marine sponges are soft-bodied, sessile organisms which belong to the Porifera phylum. In both salt and freshwater ecosystems, more than 8000 species of sponges have been described. The sessile nature of marine sponges has led to the development of mechanisms for chemical defense to deter marine predators such as sharks, tortoises, and invertebrates [7]. Marine sponges are therefore an extremely rich source of secondary metabolites possessing various biological activities. Investigation of Red Sea marine sponges permitted detection of a wide range of secondary metabolites, including terpenes, alkaloids, sterols, steroidal glycosides, and ceramides [8–10]. Ceramides are bioactive lipids, which have been found in many marine invertebrate organisms. These compounds are involved in a number of physiological functions including apoptosis, arrest of cell growth, and cell-senescence [11]. They have also been reported to be precursors of complex sphingolipids (SLs). Ceramides possessing cytotoxic activity have previously been isolated from marine sponges [12]. The marine sponge *Stylissa carteri* is abundant in coastal Red Sea reefs, typically at depths between 5 and 15 m. Numerous secondary metabolites have already been isolated from *S. carteri*, including alkaloids [13], cyclic heptapeptides [14], and the pyrrole-2-aminoimidazoles stylissazoles A-C [15]. Alkaloids isolated from *S. carteri* were suggested to be prospective supports for human immunodeficiency virus (HIV) inhibition [16]. In addition, *Stylissa carteri*-associated bacteria were reported to have anti-plasmodial activity [17]. In this study, a bioactivity-guided fractionation was performed assisted by LC-HRESIMS investigation of the Red Sea sponge *Stylissa carteri* methanolic extract, which led to the isolation of two new compounds, stylissamide A (**1**) and stylissoside A (**2**). The potential cytotoxic activities of the isolated compounds are also reported, in addition to the investigation of a possible mechanism of cytotoxic activity through molecular docking simulation studies.

2. Results and Discussion

2.1. Structure Elucidation of the Isolated Compounds

Compound **1** (Figure 1) was obtained as a white powder, and its molecular formula was determined to be $C_{32}H_{65}NNaO_5$ by HRESIMS with m/z 566.4772 $[M + Na]^+$ (calcd 566.4760), representing one degree of unsaturation (Supplementary Materials, Figure S1). The ^1H and ^{13}C NMR spectral data of compound **1** are listed in Table 1 (Supplementary Materials, Figures S2–S7). The ^1H NMR spectrum (measured in C_5D_5N, 400 MHz), displayed resonances of an amide proton doublet at δ_H 8.95 (d, J = 8.4 Hz) and a long methylene chain's protons at δ_H 1.25, representing a sphingolipid skeleton. Characteristic resonances of the hydrocarbon chain unit 2-amino-1,3,4,2'-tetrol were observed at δ_H 5.12 (m), (dd, J = 8.0, 4.8 Hz), 4.43 (dd, J = 8.0, 4.8 Hz), 4.29 (m), 4.62 (m), and 4.37 (m) assigned to H-2, H-1, H-3, H-4, and H-2', respectively. Resonances corresponding to the aliphatic methyl groups at δ_H 0.85 (t, J = 6.8 Hz) are assigned to CH_3-17 and CH_3-15'.

The ^{13}C NMR spectrum (C_5D_5N, 100 MHz) of **1** showed 32 carbon signals. Characteristic resonances of a 2-amino-1,3,4,2'-tetrol unit of the hydrocarbon chain were observed at δ_C 52.7 (C-2), 61.8 (C-1), 76.5 (C-3), 72.7 (C-4), and 72.2 (C-2'). In addition, there was a resonance at δ_C 14.5 assigned for the two terminal methyl groups (C-17 and C-15') and at δ_C 175.0 assigned for the amide carbonyl (C-1'). Analysis of the ^1H-^1H COSY, HMQC, and HMBC (Supplementary Materials, Figures S8–S10) spectra led to the assignment of all proton and carbon signals for compound **1**. The positions of the hydroxy groups were confirmed by ^1H-^1H COSY correlations between 2H-1/H-2, H-2/H-3, H3/H-4, H-4/H-5, and

H-2'/H-3' and by the HMBC correlations of 2H-1/C-2, 2H-1/C-3, H-3/C-4, H-3/C-5, H-4/C-2, H-4/C-3, H-2/C-1', and H-2'/C-1', leading to the assignment of C-1, C-2, C-3, C-4, C-1', and C-2' (Figure 2).

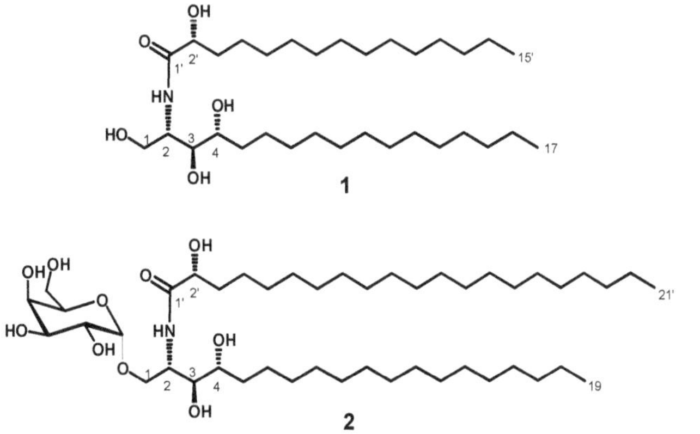

Figure 1. Chemical structures of the newly isolated compounds: stylissamide A (**1**) and stylissoside A (**2**).

Table 1. ^{1}H (400 MHz) and ^{13}C NMR (100 MHz) for the new compounds **1** and **2** in C_5D_5N.

	1			**2**	
Position	δ_H (mult., J_{Hz})	δ_C	**Position**	δ_H (mult., J_{Hz})	δ_C
1	4.43, dd (8.0, 4.8)	61.8	1a	4.32, m	68.2
2	5.12, dd (8.0, 4.8)	52.7	1b	4.59, m	
3	4.29, m	76.5	2	5.29, m	50.4
4	4.62, m	72.7	3	4.39, m	76.5
5	1.25, m	30.2	4	4.28, m	72.3
6	1.25, m	30.0	5	1.27, m	31.9
7–14	1.25, m	29.7	6	1.27, m	30.2
15	1.25, m	29.9	7-18	1.27, m	29.9
16	1.25, m	22.7	19	0.85, t (6.8)	14.2
17	0.85, t (6.8)	14.1	1'	-	175.0
1'	-	175.0	2'	4.63, m	72.4
2'	4.37, m	72.2	3'	2.00, m	31.9
3'	2.05, m	30.2	4'	1.27, m	30.2
4'	1.25, m	30.0	5'–20'	1.27, m	29.9
5'13'	1.25, m	29.7	21'	0.85, t (6.8)	14.2
14'	1.25, m	22.7	NH	8.53, d (8.4)	-
15'	0.85, t (6.8)	14.1	1''	5.61, d (3.4)	101.2
NH	8.59, d (8.4)		2''	4.64, m	70.2
			3''	4.50, m	71.6
			4''	4.54, m	71.0
			5''	4.50, m	73.1
			6''	4.33, m	62.6

The fatty acid length was determined on the basis of the results of its methanolysis followed by peak detection by HRMS, which showed a molecular ion peak at 295.2249 [M + Na]$^+$ (calcd for $C_{16}H_{32}NaO_3$, *m/z* 295.2249) indicating a C_{15} fatty acid methyl ester for compound **1**. The ceramide moieties' configuration was assigned by comparing its physical data, ^{1}H NMR and ^{13}C NMR (measured in C_5D_5N) values with those of its analogs, (likewise using pyridine) reported in the literature, wherein the optical rotation +17.4 (*c* 1.00, MeOH) and the chemical shifts of C-2 (δ 52.7), C-3 (δ 76.5), C-4 (δ 72.7), and C-2' (δ 72.2) in addition to the chemical shifts of their corresponding protons were in good agreement with those of phytosphingosine-type ceramides possessing (2*S*, 3*S*, 4*R*, and

2′R) configurations [18–20]. This evidence suggested the relative configurations of C-2, C-3, C-4, and C-2′ to be 2S, 3S, 4R, and 2′R, respectively. Accordingly, the full structure of **1** was assigned as (R)-2′-hydroxy-N-[(2S,3S,4R)-1,3,4-trihydroxyheptadecan-2-yl]pentadecanamide (stylissamide A), which, to the best of our knowledge, is a new compound.

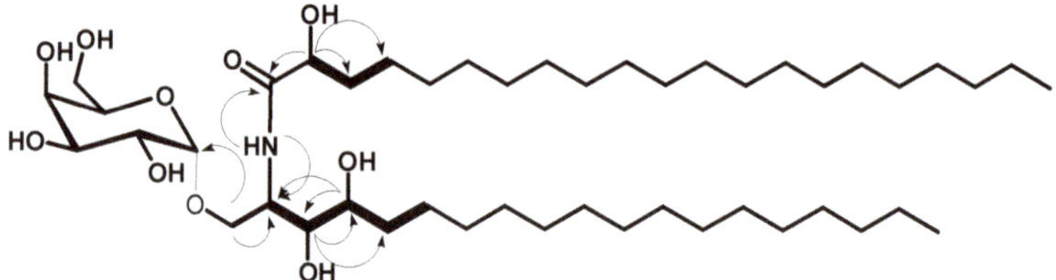

Figure 2. Key ^{1}H-^{1}H COSY (bold) and HMBC (arrows) correlations in compound **1**.

Compound **2** (Figure 1) was obtained as a white amorphous powder, and its molecular formula was determined to be $C_{46}H_{91}NNaO_{10}$ by HRESIMS at *m/z* 840.6547 [M + Na]$^+$ (calcd for *m/z* 840.6541), representing one degree of unsaturation (Supplementary Materials, Figure S11). The ^{1}H and ^{13}C NMR spectral data of compound **2** are listed in Table 1 and in the Supplementary Materials, (Figures S12–S18). The ^{1}H NMR spectrum in (C_5D_5N, 400 MHz) displayed resonances of an amide proton doublet at δ_H 8.48 (1H, d, *J* = 8.4 Hz) and a long methylene chain's protons at δ_H 1.25, representing a sphingolipid skeleton. Characteristic resonances of a 2-amino-1,3,4,2′-tetrol unit of the hydrocarbon chain were observed at δ_H 5.29 (1H, m), 4.63 (1H, m), 4.32 (1H, m), 4.59 (1H, m), 4.39 (1H, m), and 4.28 (1H, m) assigned for H-2, H-2′, H-1b, H-1a, H-3, and H-4 respectively. In addition, resonances corresponding to aliphatic methyl groups at δ_H 0.85 (3H, t, *J* = 6.8 Hz) assigned for CH_3-19 and CH_3-21′. The ^{13}C NMR spectrum in (C_5D_5N, 100 MHz), showed 46 carbon signals. Characteristic resonances of a 2-amino-1,3,4,2′-tetrol unit of the hydrocarbon chain were observed at δ_C 50.4 (C-2), 72.4 (C-2′), 68.2 (C-1), 76.5 (C-3), and 72.3 (C-4). In addition, there was a resonance at δ_C 14.2 assigned for the two terminal methyl groups (C-19 and C-21′) and at δ_C 175.0 attributed to the amide carbonyl (C-1′). The ^{13}C NMR spectrum revealed the presence of an anomeric carbon δ_C 101.2 together with the characteristic signals at δ_C 70.2, 71.6, 71.0, 73.1, and 62.6 indicating the presence of a sugar moiety. The structure of compound **2** was characterized by comparison of its ^{13}C NMR spectral data with those of the known cerebroside, agelasphin, possessing a 2-hydroxy fatty acid group [21,22]. The ^{1}H NMR signal at δ_H 5.61 (d, *J* = 3.4 Hz) clearly indicated that the galactose had an α-linkage [21]. Analysis of the ^{1}H-^{1}H COSY, HMQC, and HMBC (Supplementary Materials, Figures S19–S21) spectra led to the assignment of the proton and carbon signals for compound **2**. The positions of the hydroxy groups were confirmed by ^{1}H-^{1}H COSY correlations between 2H-1/ H-2, H-2/H-3, H-3/H-4, H-4/2H-5, and H-2′/2H-3′ and from HMBC correlations of 2H-1/C-2, 2H-1/C-3, H-3/C-4, H-3/C-5, H-4/C-2, H-4/C-3, H-2/C-1′, and H-2′/C-1′, leading to the assignment of C-1/C-2/C-3/C-4/C-1′/C-2′ (Figure 3).

Figure 3. Key ^{1}H-^{1}H COSY (bold) and HMBC (arrows) correlations for compound **2**.

In a similar way as for compound **1**, the chain length of the fatty acid was determined based on the results of its methanolysis followed by detecting peaks from HRMS. The HRMS showed one molecular ion peak at 379.3188 [M + Na]$^+$ (calcd for $C_{22}H_{44}NaO_3$: 379.3188) indicating the presence of a C_{21} fatty acid methyl ester for compound **2**. As for compound **1**, The cerebroside's relative configuration was suggested to be (2*S*, 3*S*, 4*R*, 2′*R*), since the optical rotation +17.40 (*c* 1.00, MeOH), the afore mentioned ^{1}H NMR (H-2, H-3, H-4, H-2′) and ^{13}C NMR signals (C-1, C-2, C-3, C-4, C-2′) were in good agreement with those of phytosphingosine-type cerebroside molecular species possessing (2*S*,3*S*,4*R*,2′*R*)-configuration [23,24]. So, the full structure of **2** was determined to be (*R*)-*N*-[(2*S*,3*S*,4*R*)-3,4-dihydroxy-1-{[(2*R*,3*R*,4*S*,5*S*,6*R*)-3,4,5-trihydroxy-6-(hydroxymethyl)tetrahydro-2*H*-pyran-2-yl)oxy]nonadecan-2-yl}-2′-hydroxyhenicosanamide (stylissoside A), which, to the best of our knowledge, is a new compound.

2.2. Metabolomic Profiling

Metabolomics is a fast growing technology that has effectively contributed to a number of plant-related sciences and drug discovery. The secondary metabolomes of sponges consist of widespread chemically distinct metabolites that represent the outcome of gene expressions in cells, and thus are valuable in recognizing different phenotypic traits [25]. In this context, metabolomic profiling of *Stylissa carteri* using LC–HRESIMS for dereplication purposes (Figure 4) led to the identification of a range of metabolites, mostly represented by oleanane saponins, phenolic diterpenes, and lupane triterpenes. The phytochemicals (Table 2) were tentatively identified by searching in some databases, e.g., the Dictionary of Natural Products (DNP) and METLIN [26,27].

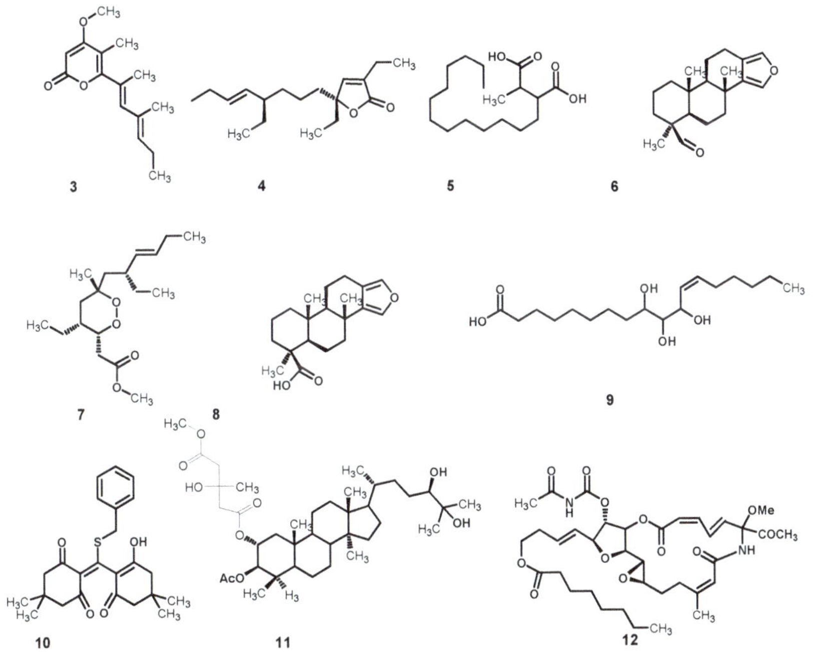

Figure 4. Chemical structures of the annotated metabolites from *Stylissa carteri*, cyercene (**3**), plakortone G (**4**), pedicellic acid (**5**), spongia-13(16),14-dien-19-al (**6**), plakortin (**7**), spongia-13(16),14-dien-19-oic acid (**8**), 9,10,11-trihydroxy-(12*Z*)-12-octadecenoic acid (**9**), benzylthiocrellidone (**10**), methyl aeruginosate C (**11**), and salarin B (**12**).

Table 2. Dereplicated metabolites from *Stylissa carteri*.

	RT (min)	MZMine ID	Molecular Weight	Name	Source	Reference
1	5.108521	209	234.1261	Cyercene (3)	Mollusca *Cyerce cristallina*	[28]
2	10.04907	57	278.2253	Plakortone G (4)	Porifera *Plakortis sp*	[29]
3	8.3637	13	314.2445	Pedicellic acid (5)	*Didymocarpus pedicellate*	[30]
4	10.10114	263	300.2095	Spongia-13(16),14-dien-19-al (6)	Porifera *Spongia officinalis*	[31]
5	7.916396	12	312.2289	Plakortin (7)	*Plakortis halichondrioides, Sponge*	[32]
6	8.382825	227	316.2043	Spongia-13(16),14-dien-19-oic acid (8)	Porifera *Spongia officinalis*	[31]
7	6.245154	20	330.2394	9,10,11-Trihydroxy-(12Z)-12-octadecenoic acid (9)	Chinese truffle *Tuber indicum*	[33]
8	3.509688	291	412.1711	Benzylthiocrellidone (10)	Porifera *Crella spinulata*	[34]
9	13.7787	225	676.4528	Methyl aeruginosate C (11)	*Stropharia aeruginosa*	[35]
10	13.4309	228	720.3484	Salarin B (12)	Porifera *Fascaplysinopsis sp*	[36]

2.3. Evaluation of the Antitumor Activity In Vitro

The potential cytotoxicity of compounds **1** and **2** isolated from *Stylissa carteri* was measured by the sulphorhodamine B (SRB) assay adopting the method of Skehan et al. [37] following the protocol described by Vichai and Kirtikara [38] on breast (MCF-7) and liver (HepG2) cancer cell lines. Color intensity was measured on an ELISA reader and the respective IC_{50} (concentration of the compound which reduces survival of cancer cells to 50%) values were calculated. As presented in Table 3, both compounds resulted in promising anticancer activities against breast (MCF-7) and hepatic (HEPG2) cancer cell lines. Compound **1** exhibited stronger cytotoxicity against MCF-7 with an IC_{50} value of 21.1 ± 0.17 µM, while compound **2** displayed a slightly lower cytotoxicity, with an IC_{50} value of 27.5 ± 0.18 µM. The case was reversed in HepG2 cancer cells, where compound **2** was more active (IC_{50} 30.5 ± 0.23 µM) than **1** (IC_{50} 36.8 ± 0.16 µM). The inhibitory properties of these compounds were compared with those of the standard drug cisplatin.

Table 3. IC_{50} values of compounds 1 and 2 on breast (MCF-7) and liver (HEPG2) cancer cell lines.

	HepG2	MCF-7
	IC_{50} (µM)	
1	36.8 ± 0.16	21.1 ± 0.17
2	30.5 ± 0.23	27.5 ± 0.18
Cisplatin	21.3 ± 0.40	15.3 ± 0.10

Each data point represents the mean ± SD of four independent experiments (significant differences at $p < 0.05$).

2.4. In Silico Studies

Modeling and docking simulations of the newly isolated compounds (**1** and **2**) were performed using the Molecular Operating Environment (MOE) software [39] and the crystal structure of Su(var)3–9, enhancer of Zeste, Trithorax (SET)/inhibitor 2 of protein phosphatase 2A (I2PP2A) oncoprotein; (PDB: 2E50) [40]. D-*erythro*(*e*)-C_{18} ceramide was chosen as a reference compound in the simulation studies due to its higher affinity for binding to SET protein compared to other endogenous ceramides or sphingosine [41]. PP2A has been reported to exhibit a tumor suppressive function by inducing apoptosis or programmed cell death in various cancerous cells [42]. Sphingolipid ceramide has long been described to activate PP2A through a direct interaction with the PP2AC catalytic domain [43], but its mechanistic details have so far remained unknown. Alternatively, another mechanism of ceramide-induced activation of PP2A has been recently found to proceed through the direct binding of ceramide with SET oncoprotein, which functions as an inhibitor of tumor suppressor PP2A [41]. Because the crystal structure of the ceramide-SET complex was not available in the protein data bank, induced fit docking was performed using the crystal structure of the SET protein and C_{18} ceramide as a flexed ligand. The top generated pose (S = −4.5237 kcal/mol) of the initial flexible docking revealed that the C_{18} ceramide chain of **1** interacts at the hydrophobic binding pocket along the space between the helix of the dimerization domain and the β sheets within the protein structure.

The pocket is predominately lipophilic with some regions of hydrophilicity (Figure 5). Furthermore, our simulation results showed that the ceramide molecule adopts an "extended" conformation inside the SET protein, in which the sphingosine and lipid chains are placed in opposite directions. This finding was further confirmed by the NMR model suggested previously by De Palma et al. [41]. The docking results of compound **1** revealed that the ligand exhibited a similar extended orientation and comparable binding interactions within the pocket ($S = -6.6821$ kcal/mol). The lipid chain of compound **1** forms hydrophobic interactions with the amino acid residues Phe-68, Tyr-127, Pro-214, and Trp-213. Additionally, along the N-terminal region, the 1,4-dihydroxy array of the dihydrosphingosine chain donates two H-bonds to Glu-111 (2.843 and 2.849 Å, respectively), while the 3-hydroxy group accepts an H-bond from Gln-65 with 2.138 Å (Figure 6A). Previous data had indicated that the amino acid residues 36–124 of the N-terminal region are involved in the inhibition of PP2A as well as the binding with ceramide [41]. On the other hand, compound **2** was found to exhibit a different compact conformation within the active site, where the dihydrosphingosine backbone runs perpendicular to both the fatty acyl chain and the sugar moiety. Therefore, compound **2** showed a weaker binding pattern within the pocket ($S = -9.2672$ kcal/mol) (Figure 6B). The lipid chain of compound **2** forms hydrophobic interactions with the amino acid residues Phe-68, Phe-67, and Pro-214. The longer chains of compound **2** destabilize the complex by shifting the ligand from the main binding sites within the SET structure. The hydroxy group of dihydrosphingosine chain donates an H-bond to the amide carbonyl group of Val-112 (2.78 Å), the 4-OH of lipid chain donates an H-bond to Asn-61 (2.77 Å), and the CH_2OH of the sugar moiety projects out the active site and donates an H-bond to Glu-116 (2.91 Å).

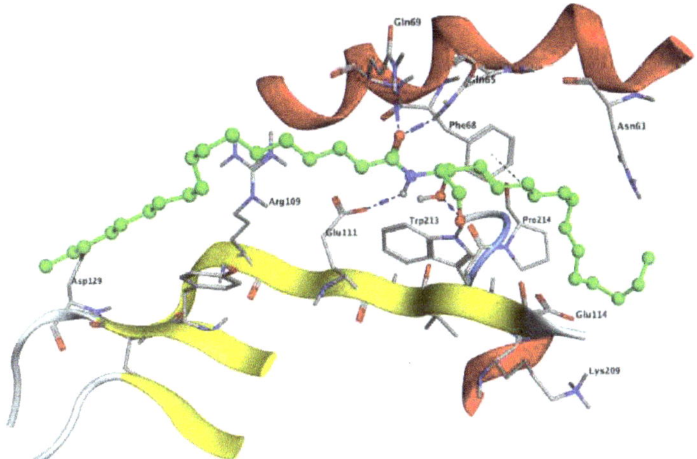

Figure 5. The top generated pose of the induced fit docking simulation oriented C_{18} ceramide (green) in an extended conformation along the space between the helix of the dimerization domain and the β sheets of the SET structure, which was further used as a generated active site for the docking simulations of compounds **1** and **2** within the SET protein.

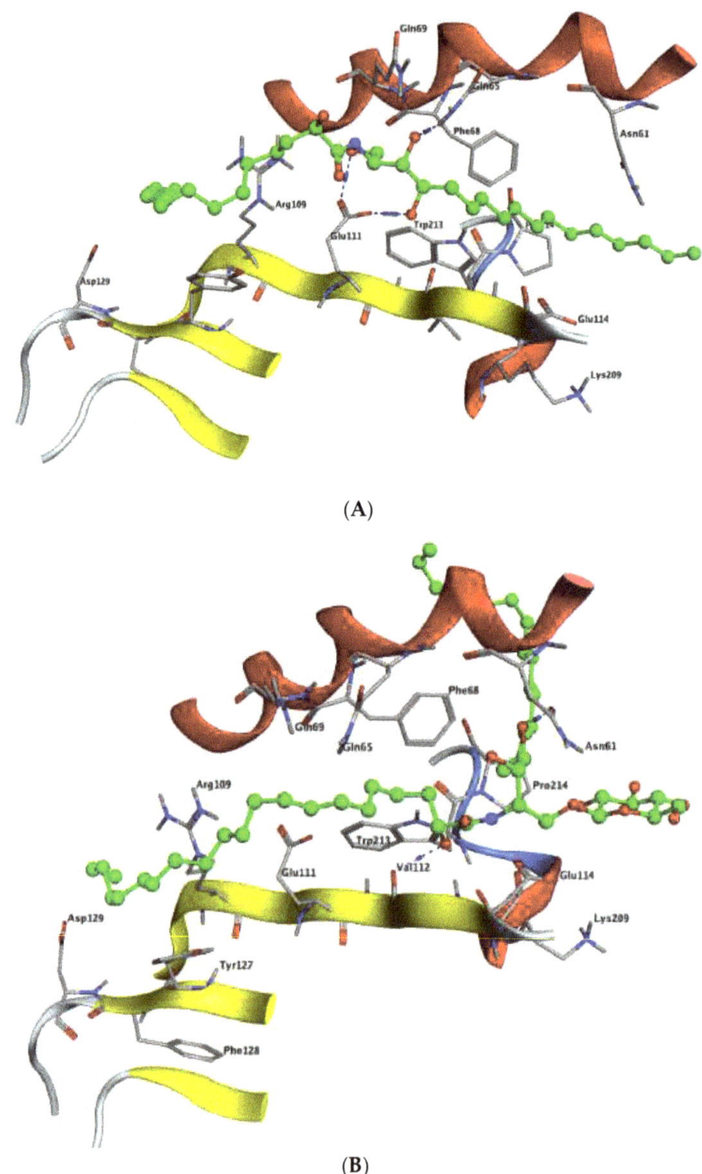

Figure 6. Docking of compounds **1** (**A**) and **2** (**B**) within the SET oncoprotein active site.

3. Materials and Methods

3.1. General Experimental Procedures

^{1}H NMR (400 MHz), ^{13}C NMR (100 MHz), DEPT-135 and 2D NMR spectra were registered on a Varian AS 400 (Varian Inc., Palo Alto, CA, USA) using the residual solvent signal as an internal standard. High-resolution mass spectra were recorded using a Bruker BioApex (Bruker Corporation) machine. Pre-coated silica gel G-25 UV254 plates were used for thin layer chromatography (TLC)

(20 cm × 20 cm) (E. Merck, Darmstadt, Germany). Silica gel (Purasil 60A, 230–400 mesh) was used for flash column chromatography (Whatman, Sanford, ME, USA).

3.2. Sponge Material

The sponge *Stylissa carteri* was collected from Sharm El Sheikh in the Egyptian Red Sea. The sponge material was air-dried and kept at low temperature ($-24\,°C$) before further processing. Identification of the sponge was performed by Dr. Tarek Temraz, Marine Science Department, Faculty of Science, Suez Canal University, Ismailia, Egypt. Voucher specimens were deposited under registration number SAA-46 in the herbarium section of Pharmacognosy Department, Faculty of Pharmacy, Suez Canal University, Ismailia, Egypt.

3.3. Extraction and Isolation

Stylissa carteri sponge (1.5 kg) was frozen, chopped to small pieces, and then extracted with methanol (3 × 2 L) at room temperature. The combined extract was concentrated in vacuo to give a reddish-brown viscous residue (30 g), which was suspended in distilled water (1 L) and partitioned with *n*-hexane, chloroform, and *n*-butanol giving rise to four fractions, namely SC-1 (4 g), SC-2 (3 g), SC-3 (17.8 g), and SC-4 (3.5 g). Fractions SC-1 and SC-2 were concentrated and combined to give SC-C (7 g), which was chromatographed on a silica gel column using $CHCl_3$: MeOH (1:0 ~ 6.5:3.5) affording eight sub-fractions (SC-C-1 - SC-C-8). Subfraction SC-C-2 (2.6 g) was chromatographed on a silica gel column using $CHCl_3$: MeOH (1:0 ~ 6.5:3.5) followed by a two-step purification on Sephadex LH-20 under isocratic conditions ($CHCl_3$:MeOH 1:1) to obtain pure compound **1** (25 mg, white amorphous powder). Subfraction SC-C-6 (450 mg) was chromatographed on a silica gel column using an isocratic elution of $CHCl_3$:MeOH (9:1 ~ 6.5:3.5) followed by a Sephadex LH-20 column using isocratic elution conditions of $CHCl_3$:MeOH (1:1) to get compound **2** (31 mg, white amorphous powder) in a pure form.

(R)-2′-Hydroxy-N-[(2S,3S,4R)-1,3,4-trihydroxyheptadecan-2-yl]pentadecanamide, stylissamide A (**1**): white, amorphous powder; ^{1}H NMR and ^{13}C NMR (see Table 1); HRESIMS (positive ion mode) *m/z* 566.4772 $[M + Na]^+$ (calcd for $C_{32}H_{65}NNaO_5$, 566.4760).

(R)-N-{(2S,3S,4R)-3,4-Dihydroxy-1-[(2R,3R,4S,5S,6R)-3,4,5-trihydroxy-6-(hydroxymethyl)tetrahydro-2H-pyran-2-yl)oxy]nonadecan-2-yl}-2′-hydroxyhenicosanamide, stylissoside A (**2**): white, amorphous powder; ^{1}H NMR and ^{13}C NMR (see Table 1); HRESIMS (positive ion mode) *m/z* 840.6547 $[M + H]^+$ (calcd for $C_{46}H_{91}NNaO_{10}$, 840.6541).

3.4. Ceramide Hydrolysis

An aliquot of 5 mg of **1** and **2** were heated with 5% HCl/MeOH (0.5 mL) at 70 °C for 8 h. The mixture was extracted with *n*-hexane and concentrated in vacuo to afford the corresponding hydroxy fatty acid methyl esters. The HRMS of **1** showed a molecular ion peak at *m/z* 295.2249 $[M + Na]^+$ (calcd for $C_{16}H_{32}NaO_3$: 295.2249) indicating the presence of a C_{15} fatty acid methyl ester. For compound **2**, the HRMS analysis showed a molecular ion peak at *m/z* 379.3188 $[M + Na]^+$ (calcd for $C_{22}H_{44}NaO_3$: 379.3188) evidencing a C_{21} fatty acid methyl ester.

3.5. Identification of the Sugar Moiety in Compound 2

Compound **2** (5 mg) was heated at 70 °C with 5% HCl/MeOH (0.5 mL) for 8 h then the reaction mixture (including the esterified fatty acid) was extracted with chloroform. The methanolic layer (containing the sugar moiety) was neutralized with Ag_2CO_3. The resulting precipitates were filtered off and the filtrate was concentrated in vacuo then injected on HPLC (Cosmosil Sugar-D, 4.6 ID × 250 mm, 1 mL/min, RI detector, 95% aqueous acetonitrile). HPLC runs of standard glucose and galactose were carried out in parallel and the respective retention times were compared with that of the unknown sugar in **2**. Galactose showed a retention time of 4.76 min, similar to that of the sugar in compound **2**, however, glucose displayed a retention time of 4.62 min, which was eluted earlier from the sugar in **2**. Therefore, the sugar in the new compound **2** was identified as galactose.

3.6. Determination of the Configuration of the Sugar Moiety in **2**

Compound **2** (2 mg) was hydrolyzed by heating in 0.5 M aqueous HCl (0.1 mL) and then neutralized with Amberlite IRA400. After drying in vacuo, the residue was dissolved in pyridine (0.1 mL) containing L-cysteine methyl ester hydrochloride (0.5 mg) and heated at 60 °C for 1 h in 0.1 mL solution of *o*-tolyl isothiocyanate (0.5 mg) in pyridine. The reaction mixture was directly analyzed by reversed-phase HPLC (Cosmosil-5-C_{18}-AR-II, 4.6 ID × 250 mm, 0.8 mL/min, λ = 250 nm, 25% acetonitrile in 50 mM H_3PO_4). The peaks at 19.58 min and 17.52 min coincided with those of the respective derivatives of D-galactose. The same derivatization procedures were performed on 5 mg of the standard compounds, D-galactose (t_R = 18.6 min) and L-galactose (t_R = 19.3 min).

3.7. Metabolomic Profiling

The metabolomics profiling was performed according to a method reported by Elsayed et al. [44]. These files were imported to the data mining software MZmine 2.10 for peak picking, deconvolution, deisotoping, alignment, and formula prediction. Dereplication of compounds was carried out by comparison with those in the Marinlit database and the Dictionary of Natural Products (DNP) 2015.

3.8. Cytotoxicity Assays

The cytotoxicity of compounds **1** and **2** was measured by the sulphorhodamine B (SRB) assay as described by Skehan [37], following the protocol described by Vichai and Kirtikara [38] on breast (MCF-7) and liver (HepG2) cancer cell lines. At an initial concentration of 3×10^3 cell/well in a 150 μL fresh medium, cells were seeded in 96-well microtiter plates, and left for 24 h to adhere to the plates. Different concentrations of the drug were added 0, 5, 12.5, 25, and 50 μg/mL respectively.

The plates were incubated for 48 h. The cells were fixed at 4 °C with 50 μL cold trichloroacetic acid (10% final concentration) for 1 h. The plates were washed with distilled water (automatic washer Tecan, Germany) and stained with 50 μL 0.4% SRB dissolved in 1% acetic acid at room temperature for 30 minutes. With 1% acetic acid, the plates were washed and air-dried. 100 μL/well of 10 M tris base (pH 10.5) was used to solubilize the dye. Using an ELISA microplate reader (Sunrise Tecan reader, Germany), the optical density of each well was measured at 570 nm spectrophotometrically. The mean background absorbance was automatically subtracted, and mean values were determined for each concentration of drugs. The experiment was repeated three times, then the IC_{50} values were calculated.

3.9. In Silico Studies

All the molecular modeling calculations and docking simulation studies were performed using the Molecular Operating Environment (MOE 2014.0901, 2014; Chemical Computing Group, Canada) software. All MOE minimizations were performed until an RMSD gradient of 0.01 kcal/mol/Å with the force field (Amber10:ETH) and gas phase solvation to calculate the partial charges automatically. Before simulations, the protein was protonated using the LigX function and the monomer was identified. Induced fit docking simulation was performed initially to predict the active site using C_{18} ceramide as a flexed ligand. Triangle matching with London dG scoring was chosen for initial placement, then the top 30 poses were refined using force field (Amber10:ETH) and Affinity DG scoring. The top pose from this simulation was analyzed and further used as a reference for the docking simulation of the newly isolated compounds **1** and **2**. The output database dock file was created with different poses for each ligand and arranged according to the final score function (S), which is the score of the last stage that was not set to zero.

4. Conclusions

The present research highlighted the efficacy of LCMS profiling when combined with bioassay-guided drug discovery from marine invertebrates to accelerate the conventionally long processes of identifying an active metabolite by successive isolation from crude extracts. Dereplication experiments focused on the chemotaxonomic sorting helped in the identification of putative active metabolites, while structural assignment of the isolated compounds, using both HRMS and NMR, supported the identified hits. Metabolomic profiling of the bioactive extract displayed the existence of numerous secondary metabolites, mainly oleanane saponins and phenolic and lupane di- and triterpenes. Therefore, bio-guided isolation combined with LC-MS metabolomic profiling of the Red Sea sponge *Stylissa carteri* crude extract was performed, leading to the characterization of two new compounds, stylissamide A (**1**) and stylissoside A (**2**), which, to the best of our knowledge, have not been isolated from any natural source before. Both compounds, showed promising cytotoxic activity against breast (MCF-7) and liver (HepG2) cancer cell lines. Compound **1** exhibited stronger cytotoxicity against breast (MCF-7) cancer cells, while compound **2** exhibited higher activity against the HepG2 cancer cell line compared to cisplatin as the standard. Moreover, a docking study was performed to investigate the possible mechanism(s) of the cytotoxic activity of **1** and **2**. The newly isolated metabolites were docked into the crystal structure of the SET oncoprotein and inhibitor 2 of protein phosphatase 2A (I2PP2A). We believe that the in vivo biological investigations of these metabolites will be of value for future anti-cancer drug development.

Supplementary Materials: The following are available online at http://www.mdpi.com/1660-3397/18/5/241/s1, Figures S1–S10: HRMS, ^{1}H NMR, ^{13}C NMR, and COSY, HMBC, and HMQC of Compound 1, Figures S11–S21: HRMS, ^{1}H NMR, ^{13}C NMR, and COSY, HMBC, and HMQC of Compound 2.

Author Contributions: Conceptualization U.R.A., S.A.A., H.A.H. and G.B.; methodology, N.A.E., E.S.H., R.F.A.A., S.F. and A.K.I.; data curation, A.M.H., A.F.M., K.Y., F.A.B., M.M.A.-S. and S.I.A.; original draft preparation, N.A.E. and U.R.A.; writing, review, and editing, all authors. All authors have read and agreed to the published version of the manuscript.

Funding: This research received no external funding.

Acknowledgments: The authors are grateful to Tarek Temraz (Marine Science Department, Faculty of Science, Suez Canal University, Ismailia, Egypt). Many thanks and appreciation are due to the Egyptian Environmental Affairs Agency (EEAA) for facilitating sample collection along the coasts of the Red Sea. S. F. thanks the German Academic Exchange Service (Deutscher Akademischer Austauschdienst, DAAD) for a generous scholarship grant.

Conflicts of Interest: The authors declare there is no conflict of interest.

References

1. Fu, Y.; Luo, J.; Qin, J.; Yang, M. Screening techniques for the identification of bioactive compounds in natural products. *J. Pharm. Biomed. Anal.* **2019**, *168*, 189–200. [CrossRef]

2. Abdelhameed, R.F.; Ibrahim, A.K.; Temraz, T.A.; Yamada, K.; Ahmed, S.A. Chemical and biological investigation of the Red Sea sponge Echinoclathria species. *Int. J. Pharm. Sci. Res.* **2017**, *9*, 1324–1328.

3. Liu, M.; El-Hossary, E.M.; Oelschlaeger, T.A.; Donia, M.S.; Quinn, R.J.; Abdelmohsen, U.R. Potential of marine natural products against drug-resistant bacterial infections. *Lancet Infect. Dis.* **2019**, *19*, 237–245. [CrossRef]

4. Hifnawy, M.S.; Aboseada, M.A.; Hassan, H.M.; Tohamy, A.F.; Abdel-Kawi, S.H.; Rateb, M.E.; El Naggar, E.B.; Quinn, R.J.; Abdelmohsen, U.R. Testicular caspase-3 and β-Catenin regulators predicted via comparative metabolomics and docking studies. *Metabolites* **2020**, *10*, 31. [CrossRef] [PubMed]

5. Abdelmohsen, U.R.; Balasubramanian, S.; Oelschlaeger, T.A.; Grkovic, T.; Pham, N.B.; Quinn, R.J.; Hentschel, U. Potential of marine natural products against drug-resistant pathogens. *Lancet Infect. Dis.* **2017**, *17*, 30–40. [CrossRef]

6. Khalifa, S.A.M.; Elias, N.; Farag, M.A.; Chen, L.; Saeed, A.; Hegazy, M.E.F.; Moustafa, M.S.; Abd El-Wahed, A.; Al-Mousawi, S.M.; Musharraf, S.G.; et al. Marine natural products: A source of novel anticancer drugs. *Mar. Drugs* **2019**, *17*, 491. [CrossRef] [PubMed]

7. Anjum, K.; Abbas, S.Q.; Shah, S.A.; Akhter, N.; Batool, S.; Shams ul-Hassan, S. Marine sponges as a drug treasure. *Biomol. Ther.* **2016**, *24*, 347–362. [CrossRef]

8. Sayed, A.M.; Alhadrami, H.A.; El-Hawary, S.S.; Mohammed, R.; Hassan, H.M.; Rateb, M.; Abdelmohsen, U.R.; Bakeer, W. Discovery of two brominated oxindole alkaloids as Staphylococcal DNA gyrase and pyruvate kinase inhibitors via inverse virtual screening. *Microorganisms* **2020**, *8*, 293. [CrossRef]

9.   El-Hawary, S.S.; Sayed, A.M.; Mohammed, M.; Hassan, H.M.; Rateb, M.E.; Amin, E.; Mohammed, T.A.; El-Mesery, M.; Abdullatif Bin Muhsinah, A.; Alsayari, A.; et al. Bioactive brominated oxindole alkaloids from the Red Sea sponge Callyspongia siphonella. *Mar. Drugs* **2019**, *17*, 465. [CrossRef]

10. Eltahawy, N.A.; Ibrahim, A.K.; Radwan, M.M.; Zayton, S.; Gomaa, M.; ElSohly, M.A.; Hassanean, H.A.; Ahmed, S.A. Mechanism of action of antiepileptic ceramide from Red Sea soft coral Sarcophyton auritum. *Bioorg. Med. Chem. Lett.* **2015**, *25*, 5819–5824. [CrossRef]

11. Hannun, Y.A.; Obeid, L.M. Principles of bioactive lipid signaling: Lessons from sphingolipids. *Nat. Rev. Mol. Cell Biol.* **2008**, *9*, 139–150. [CrossRef] [PubMed]

12. Eltahawy, N.A.; Ibrahim, A.K.; Gomaa, M.S.; Zaitone, S.A.; Radwan, M.M.; Hassanean, H.A.; ElSohly, M.A.; Ahmed, S.A. Anxiolytic and anticonvulsant activity followed by molecular docking study of ceramides from the Red Sea sponge Negombata sp. *Med. Chem. Res.* **2019**, *28*, 1818–1827. [CrossRef]

13. Giles, E.C.; Saenz-Agudelo, P.; Berumen, M.L.; Ravasi, T. Novel polymorphic microsatellite markers developed for a common reef sponge Stylissa carteri. *Mar. Biodiv.* **2013**, *43*, 237–241. [CrossRef]

14. Linington, R.G.; Williams, D.E.; Tahir, A.; Van Soest, R.; Andersen, R.J. Latonduines A and B, new alkaloids isolated from the marine sponge Stylissa carteri: Structure elucidation, synthesis, and biogenetic implications. *Org. Lett.* **2003**, *15*, 2735–2738. [CrossRef] [PubMed]

15. Patel, K.; Laville, R.; Martin, M.; Tilvi, S.; Moriou, C.; Gallard, J.; Ermolenko, L.; Debitus, C.; Al-Mourabit, A. Unprecedented stylissazoles A–C from *Stylissa carteri*: Another dimension for marine pyrrole-2-aminoimidazole metabolite diversity. *Angew. Chem. Int. Ed.* **2010**, *49*, 4775–4779. [CrossRef] [PubMed]

16. O'Rourke, A.; Kremb, S.; Bader, T.; Helfer, M.; Schmitt-Kopplin, P.; Gerwick, W.; Brack-Werner, R.; Voolstra, C. Alkaloids from the sponge *Stylissa carteri* present prospective scaffolds for the inhibition of Human Immunodeficiency Virus 1 (HIV-1). *Mar. Drugs* **2016**, *14*, 28. [CrossRef]

17. Inbaneson, S.J.; Ravikumar, S. In vitro antiplasmodial activity of marine sponge *Stylissa carteri* associated bacteria against Plasmodium falciparum. *Asian Pac. J. Trop. Dis.* **2012**, *2*, 370–374. [CrossRef]

18. Sun, Y.; Xu, Y.; Liu, K.; Hua, H.; Zhu, H.; Pei, Y. Gracilarioside and gracilamides from the red alga Gracilaria asiatica. *J. Nat. Prod.* **2006**, *69*, 1488–1491. [CrossRef]

19. Azuma, H.; Takao, R.; Niiro, H.; Shikata, K.; Tamagaki, S.; Tachibana, T.; Ogino, K. Total syntheses of symbioramide derivatives from L-Serine and their antileukemic activities. *J. Org. Chem.* **2003**, *68*, 2790–2797. [CrossRef]

20. Sandjo, L.; Hannewald, P.; Yemloul, M.; Kirsch, G.; Ngadjui, B. Triumfettamide and Triumfettoside Ic, two ceramides and other secondary metabolites from the stems of wild Triumfetta cordifolia A. Rich. (Tiliaceae). *Helv. Chim. Acta.* **2008**, *91*, 1326–1335. [CrossRef]

21. Natori, T.; Morita, M.; Akimoto, K.; Koezuka, Y. Agelasphins, novel antitumor and immunostimulatory cerebrosides from the marine sponge Agelas mauritianus. *Tetrahedron Lett.* **1994**, *50*, 2771–2784. [CrossRef]

22. Natori, T.; Koczuka, Y.; Higa, T. Agelasphins, novel alpha-galactosylceramides from the marine sponge Agelas mauritianus. *Tetrahedron Lett.* **1993**, *34*, 5591–5592. [CrossRef]

23. Kawatake, S.; Nakamura, K.; Inagaki, M.; Higushi, R. Isolation and structural determination of six glucocerebrosides from the starfish Luidia maculata. *Chem. Pharm. Bull.* **2002**, *50*, 1091–1096. [CrossRef] [PubMed]

24. Chen, X.; Wu, Y.; Chen, D. Structure determination and synthesis of a new cerebroside isolated from the traditional Chinese medicine Typhonium giganteum. *Engl. Tetrahedron Lett.* **2002**, *43*, 3529–3532. [CrossRef]

25. Abdelhafez, O.H.; Othman, E.M.; Fahim, J.R.; Desoukey, S.Y.; Pimentel-Elardo, S.M.; Nodwell, J.R.; Schirmeister, T.; Tawfike, A.; Abdelmohsen, U.R. Metabolomics analysis and biological investigation of three Malvaceae plants. *Phytochem. Anal.* **2019**, *31*, 204–214. [CrossRef] [PubMed]

26. Dictionary of Natural Products. Available online: http://dnp.chemnetbase.com/faces/chemical/ChemicalSearch.xhtml (accessed on 28 August 2017).

27. METLIN. Available online: http://metlin.scripps.edu/index.php (accessed on 15 September 2017).

28. Vardaro, R.R.; Matzo, V.D.; Crispino, A.; Cimino, G. Cyercenes, novel polypropionate pyrones from the autotomizing Mediterranean mollusc Cyerce cristallina. *Tetrahedron Lett.* **1991**, *41*, 5569–5576. [CrossRef]

29. Gochfeld, D.J.; Hamann, M.T. Isolation and biological evaluation of filiformin, plakortide F, and plakortone G from the Caribbean sponge Plakortis sp. *J. Nat. Prod.* **2001**, *64*, 1477–1479. [CrossRef]

30. Rao, K.V.; Seshadri, T.R.; Sood, M.S. Isolation and constitution of pedicellic acid: A new dicarboxylic acid from the leaves of Didymocarpus pedicellata. *Tetrahedron Lett.* **1966**, *22*, 1495–1498. [CrossRef]

31. Li, C.; Schmitz, F.J.; Kelly-Borges, M. Six new spongian diterpenes from the sponge Spongia matamata. *J. Nat. Prod.* **1999**, *62*, 287–290. [CrossRef]

32. Martin, D.; Higgs, D.; Faulkner, J. Plakortin, an antibiotic from Plakortis halichondrioides. *J. Org. Chem.* **1978**, *43*, 3454–3457.

33. Gao, J.; Wang, C.; Zhang, A.; Liu, J. A new trihydroxy fatty acid from the ascomycete, Chinese truffle Tuber indicum. *Lipids* **2001**, *36*, 1365–1370. [CrossRef] [PubMed]

34. Pattenden, G.; Wickramasinghe, W.A.; Bandaranayake, W.M. Benzylthiocrellidone, a novel thioether with strong UV A and B absorption from the Great Barrier Reef sponge Crella spinulata (Poecilosclerida: Crellidae). *Article* **2002**, *9*, 205–216.

35. Shiono, Y.; Sugasawa, H.; Kurihara, N.; Nazarova, M.; Murayama, T.; Takahashi, K.; Ikeda, M. Three lanostane triterpenoids from the fruiting bodies of Stropharia aeruginosa. *J. Asian Nat. Prod. Res.* **2005**, *7*, 735–740. [CrossRef] [PubMed]

36. Aknin, M.; Gros, E.; Vacelet, J.; Kashman, Y.; Gauvin-Bialecki, A. Sterols from the Madagascar sponge Fascaplysinopsis sp. *Mar. Drugs* **2010**, *8*, 2961–2975. [CrossRef] [PubMed]

37. Skehan, P.; Storeng, R.; Scudiero, D.; Monks, A.; McMahn, J.M.; Vistica, D.; Warren, J.; Bokesch, H.; Kenney, S.; Boyd, M.R. New colorimetric cytotoxicity assay for anticancer-drug screening. *J. Nat. Cancer Inst.* **1990**, *82*, 1107–1112. [CrossRef]

38. Vichai, V.; Kirtikara, K. Sulforhodamine B colorimetric assay for cytotoxicity screening. *Nat. Protoc.* **2006**, *1*, 1112–1116. [CrossRef]

39. Chemical Computing Group Inc. *Molecular Operating Environment (MOE) 2014.0901*; Chemical Computing Group Inc.: Montreal, QC, Canada, 2016.

40. Muto, S.; Senda, M.; Akai, Y.; Sato, L.; Suzuki, T.; Nagai, R.; Senda, T.; Horikoshi, M. Relationship between the structure of SET/TAF-Iβ/INHAT and its histone chaperone activity. *Proc Natl Acad Sci. USA* **2007**, *104*, 4285–4290. [CrossRef]

41. De Palma, R.M.; Parnham, S.R.; Li, Y.; Oaks, J.J.; Peterson, Y.K.; Szulc, Z.M.; Roth, B.M.; Xing, Y.; Ogretmen, B. The NMR-based characterization of the FTY720-SET complex reveals an alternative mechanism for the attenuation of the inhibitory SET-PP2A interaction. *FASEB J.* **2019**, *33*, 7647–7666. [CrossRef]

42. Liu, J.; Beckman, B.S.; Foroozesh, M. A review of ceramide analogs as potential anticancer agents. *Future Med. Chem.* **2013**, *5*, 1405–1421. [CrossRef]

43. Mullen, T.D.; Obeid, L.M. Ceramide and apoptosis: Exploring the enigmatic connections between sphingolipid metabolism and programmed cell death. *Anticancer. Agents Med. Chem.* **2012**, *12*, 340–363. [CrossRef]

44. Elsayed, Y.; Refaat, J.; Abdelmohsen, U.R.; Othman, E.M.; Stopper, H.; Fouad, M.A. Metabolomic profiling and biological investigation of the marine sponge-derived bacterium Rhodococcus sp. UA13. *Phytochem. Anal.* **2018**, *29*, 543–548. [CrossRef] [PubMed]